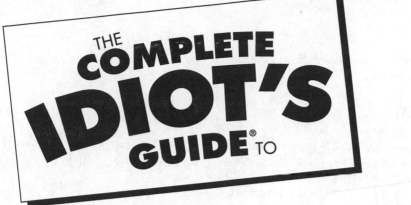

THE
COMPLETE
IDIOT'S
GUIDE® TO

Anatomy

Illustrated

by Mark F. Seifert, Ph.D.

ALPHA

A member of Penguin Group (USA) Inc.

To Linda, my wife, and my children, Laura and Mike, for their constant love, patient support, and encouragement.

ALPHA BOOKS

Published by the Penguin Group

Penguin Group (USA) Inc., 375 Hudson Street, New York, New York 10014, USA

Penguin Group (Canada), 90 Eglinton Avenue East, Suite 700, Toronto, Ontario M4P 2Y3, Canada (a division of Pearson Penguin Canada Inc.)

Penguin Books Ltd, 80 Strand, London WC2R 0RL, England

Penguin Ireland, 25 St. Stephen's Green, Dublin 2, Ireland (a division of Penguin Books Ltd.)

Penguin Group (Australia), 250 Camberwell Road, Camberwell, Victoria 3124, Australia (a division of Pearson Australia Group Pty. Ltd.)

Penguin Books India Pvt. Ltd., 11 Community Centre, Panchsheel Park, New Delhi—110 017, India

Penguin Group (NZ), 67 Apollo Drive, Rosedale, North Shore, Auckland 1311, New Zealand (a division of Pearson New Zealand Ltd.)

Penguin Books (South Africa) (Pty.) Ltd, 24 Sturdee Avenue, Rosebank, Johannesburg 2196, South Africa

Penguin Books Ltd., Registered Offices: 80 Strand, London WC2R 0RL, England

Copyright © 2008 by Mark F. Seifert, Ph.D.

International Standard Book Number: 978-1-59257-760-6
Library of Congress Catalog Card Number: 2008920825

10 09 08   8 7 6 5 4 3 2 1

Interpretation of the printing code: The rightmost number of the first series of numbers is the year of the book's printing; the rightmost number of the second series of numbers is the number of the book's printing. For example, a printing code of 08-1 shows that the first printing occurred in 2008.

Printed in the United States of America

Note: This publication contains the opinions and ideas of its author. It is intended to provide helpful and informative material on the subject matter covered. It is sold with the understanding that the author and publisher are not engaged in rendering professional services in the book. If the reader requires personal assistance or advice, a competent professional should be consulted.

The author and publisher specifically disclaim any responsibility for any liability, loss, or risk, personal or otherwise, which is incurred as a consequence, directly or indirectly, of the use and application of any of the contents of this book.

Most Alpha books are available at special quantity discounts for bulk purchases for sales promotions, premiums, fundraising, or educational use. Special books, or book excerpts, can also be created to fit specific needs.

For details, write: Special Markets, Alpha Books, 375 Hudson Street, New York, NY 10014.

Publisher: *Marie Butler-Knight*
Editorial Director: *Mike Sanders*
Senior Managing Editor: *Billy Fields*
Acquisitions Editor: *Tom Stevens*
Development Editor: *Julie Bess*
Senior Production Editor: *Jan Lynn*
Copy Editor: *Mike Dietsch*

Cartoonist: *Steve Barr*
Cover Designer: *Bill Thomas*
Book Designer: *Trina Wurst*
Indexer: *Brad Herriman*
Layout: *Chad Dressler*
Proofreader: *Mary Hunt*

Contents at a Glance

Part 1:   **Getting a Lay-of-the-Land View**   **3**

1   What Is Human Anatomy?   5
Learn a bit about the history and purpose of the study of the human body.

2   Cut from the Same Cloth: The Four Types of Tissue   11
When cells get together, they form tissue. Here you learn about the different types of tissue in the body.

Part 2:   **All Systems Go!: Organs and Body Systems**   **25**

3   Only Skin Deep: The Integumentary System   27
The skin is the body's biggest organ. Learn how it's made and what it can do for you.

4   Skeleton Crew: The Skeletal System   35
No bones about it—without a skeleton, your life would be at a standstill. Here are the hard facts about this vital system.

5   Grand Central Station: The Central Nervous System   45
The CNS controls every function and sensation in the body. Here's how it works.

6   Left of Center: The Peripheral Nervous System   61
The peripheral nerves relay messages between the CNS and the rest of the body. Learn about their intricate structure and function.

7   The Heart of the Matter: The Cardiovascular System   73
From the heart to distant tiny capillaries, the cardiovascular system pumps oxygen and nutrients throughout the body.

8   Low Fluid Level: The Lymphatic System   89
The lymph system collects excess fluid and protects the body from foreign invaders. Learn the parts of this unsung bodily hero.

9   You Take My Breath Away: The Respiratory System   97
The respiratory system begins at the nose and ends at the lungs. It takes in oxygen and helps get it to all parts of the body.

10   The Long, Winding Food Tube: The Digestive System   109
Several organs play a role in getting the nutrients from your food and helping the body use them as fuel. And when all is said and done, this system also gets rid of the undigested waste.

11   Making Water: The Urinary System   123
When you gotta go, you gotta go. Here's how the body processes fluids and rids itself of metabolic waste products.

12 The Male Reproductive System   131
    *Learn the parts of the male reproductive system and how sperm is produced
    and delivered.*

13 The Female Reproductive System   139
    *Scene of the miracle of life, the female reproductive system is a complex
    team of organs.*

Part 3:   Putting It All Together: Regional Context   149

14 The Language of Anatomy: Planes, Position, and Movement   151
    Learn the lingo of anatomists.

15 Rear View: The Back   159
    The parts and structures that bring up the rear.

16 The Upper Limb: From the Waist Up   177
    *The upper limb is made up of the shoulder, arm, forearm, and hand. These
    muscle-powered bony levers give us great independence and functional capa-
    bilities.*

17 The Pectoral Region and Thorax: The Chest Area   201
    *The thoracic cavity is home to the pleural cavities and lungs and the peri-
    cardial cavity and the heart.*

18 The Abdomen: The Belly of the Beast   215
    *The abdomen contains the stomach, the small and large intestines, the liver,
    the spleen, the pancreas, the adrenal glands, and the kidneys.*

19 The Pelvis and Perineum: Stuck in the Middle of You   227
    *The pelvis serves two main purposes: to support and protect the pelvic
    organs, and to transfer the weight of the body to the lower limbs.*

20 The Lower Limb: From the Waist Down   239
    *The lower limb consists of the hip, thigh, leg, and foot. It supports the
    weight of the body and serves as its transportation.*

21 "On Top of Old Shoulders": The Head and Neck   257
    This area is the headquarters of the body's five senses.

Appendixes

A Glossary   277

B Further Reading   291

   Index   295

Contents

Part 1:   Getting a Lay-of-the-Land View   **3**

1   What Is Human Anatomy? 5
     A Brief History of an Old Subject 5
         Early Greeks: Herophilus and Erasistratus 6
         Physician to Gladiators and Dissector of Pigs: Galen 6
         The Modern Father of Anatomy: Vesalius 7
         Under the Microscope: Leeuwenhoek 7
     The "ologies" of Human Anatomy 7

2   Cut from the Same Cloth: The Four Types of Tissue 11
     Epithelial Tissue: The Great Shoreline of the Body 11
         Epithelia on a Need-to-Know Basis 12
         The Different Types of Epithelial Tissue 12
     The Tie That Binds: Connective Tissue 14
         Connective Tissue Cells 14
         Connective Tissue Fibers 15
         Ground Substance 16
         Types of Connective Tissue 16
     Pump Me Up: Muscular Tissue 17
         Skeletal Muscle 18
         Cardiac Muscle 18
         Smooth Muscle 19
     Nervous in the Service: Nervous Tissue 19
         Neurons 20
         Glial Cells 22

Part 2:   All Systems Go!: Organs and Body Systems   **25**

3   Only Skin Deep: The Integumentary System 27
     The Skinny on Skin 28
         Epidermis: Where the Beauty Is 28
         The Dermis: What Lies Beneath 30
     Things Made from the Epidermis 31
         Hairs and Nails 31
         Glands 32

4   Skeleton Crew: The Skeletal System 35
     Bone Basics 35
        Call Me on My Cell *36*
        Bone Growth and Development *36*
     The Two Divisions of the Skeletal System 39
        The Axial Skeleton: Skull, Spine, Ribs, and Breastbone *39*
        The Appendicular Skeleton: Everything Else *41*
     Jumpin' Joints: Where Two Bones Come Together 41
        Nonsynovial Joints *41*
        Synovial Joints *42*

5   Grand Central Station: The Central Nervous System 45
     The Brain Trust: Parts of the Brain 45
        Cerebrum: Where the Real Thinking Happens *47*
        The Diencephalon *51*
        Cerebellum: Chewing Gum and Walking at the Same Time *52*
        The Brain Stem: Automatic Pilot *52*
     Spinal Cord: The Information Superhighway 54
     Meninges: They've Got You Covered 55
     The Ventricular System and Cerebrospinal Fluid: Holes and the Stuff
        That Fills Them 58

6   Left of Center: The Peripheral Nervous System 61
     Full of Nerves: The Many Aspects of the PNS 61
     Spinal Nerves: The Messengers 63
        The Somatic Component of Spinal Nerves *64*
        The Autonomic Component of Spinal Nerves *65*
     Cranial Nerves: The Special Sensory Connection 67
     It's About Regulation 68
        The Somatic Nervous System: The Outsiders *68*
        The Autonomic Nervous System: Fight or Flight vs. Rest and Digest *69*

7   The Heart of the Matter: The Cardiovascular System 73
     External Features of the Heart: It's a Cone Thing 73
        The Groove Is in the Heart *74*
        The Sender: Coronary Arteries *75*
        Return to Sender: Cardiac Veins *77*
     Internal Features of the Heart: I Have a Partition of Four! 78
        Chambers of Commerce *78*
        Wall-to-Wall Layers *80*
        Heart Valves: An Open-and-Shut Case *80*

The Great Vessels and Beyond...82
  The Aorta and Major Arteries to the Body: Systemic Suppliers *83*
  The Anatomy of a Blood Vessel...*84*
  The Very Capable Capillaries...*85*
Return Trip Home: The Vein Family...86
  You're So Vena Cava *86*
  The Portal Venous System: The Doors to Your Digestive Organs *87*

8   Low Fluid Level: The Lymphatic System 89
      Where It All Begins: The Lymphatic Capillaries.......................................89
      In-line Filters: Lymph Nodes..90
      Where the Lymphatic Action Is...91
        Lymph Trunks *91*
        Thoracic Duct..*92*
      Lymphoid Organs: The Pit Bulls of the Immune System 92
        Lymph Nodes *93*
        The Spleen *94*
        The Thymus...*94*

9   You Take My Breath Away: The Respiratory System.................................97
      Look Out Below: The Upper Respiratory Tree 97
        Who Nose?: The Nasal Cavity..*98*
        Mouth (Oral Cavity): Take a Deep Breath *99*
        Pharynx: The Throat *99*
      Larynx: The Voice Box...100
      The Lower Respiratory Tree..102
        Trachea and Bronchi: The Trunk and Branches..*103*
        Lungs: The Leaves of the Respiratory Tree *104*
      In with the New, Out with the Old: Gaseous Exchange...........................107

10   The Long, Winding Food Tube: The Digestive System 109
       Chew on This! The Mouth..109
         The Tongue *110*
         Salivary Glands *110*
       Down the Hatch: The Pharynx and Esophagus.......................................111
       Pass Me an Antacid: The Stomach 112
         Roadmap of the Stomach...*113*
         Ridges, Pits, Glands...*113*
         Layers, Muscles, and Sphincters *113*
       Fully Absorbed: The Small Intestine 114
         The Duodenum *114*
         The Jejunum...*117*
         The Ileum *117*

The End Is in Sight! The Large Intestine 118
Thanks, We Needed That! Accessory Organs of Digestion 118
    The Pancreas 119
    The Liver 119
    The Gallbladder 121

11   Making Water: The Urinary System 123
    You're In: The Kidneys 123
    Have You Met My Nephron? 126
    Ureters: Shortcut from the Kidneys to the Bladder 127
    Waiting to Go: The Urinary Bladder 128
    Ahh, Sweet Relief: The Urethra 128

12   The Male Reproductive System 131
    Testes: One, Two… 131
        Sex Hormone Production 133
    The Epididymis and Ductus Deferens 133
    Accessorize Your Genitals: The Accessory Glands 135
        Seminal Vesicles 135
        The Prostate Gland 135
        The Bulbourethral Glands 136
    The Penis 136

13   The Female Reproductive System 139
    Internal Reproductive Organs 139
        Madame Ovary 140
        Catch and Release: The Uterine Tubes 141
        The Uterus: The Cradle of Life 142
        The Vagina 144
    External Genitalia 144
    Thanks for the Mammaries! 145

Part 3:   **Putting It All Together: Regional Context**   **149**

14   The Language of Anatomy: Planes, Position, and Movement 151
    Not Just Any Position: The Anatomical Position 151
    On a Different Plane 152
    Location, Location, Location: Anatomical Terms of Relationship
      or Position 152
    And, Action!: Anatomical Terms of Movement 154

15   Rear View: The Back 159

    The Spine: Backbone of the Human Body (Literally) 159

        Nice Curves! Normal Curves of the Spine *160*

        Throw Me a Curve! Abnormal Curves *160*

    Are You from Around These Parts? Parts of a Typical Vertebra 163

    What Region Are You From? Regional Characteristics of Vertebrae 164

        Can We "Neck"? Cervical Vertebrae *164*

        Thoracic Vertebrae: This Will Stick to Your Ribs *166*

        Lumbar Vertebrae (a Pain in the Lower Back) *166*

        Back to Back, Sacroiliac: The Sacrum *167*

        Coccyx: Not a Prehistoric Bird *169*

    Action at the Vertebral Joint 169

    Ligaments of the Vertebral Column 170

    A Class of Their Own: Deep Back Muscles 172

    Spinal Cord, Spinal Nerves, and Meninges 173

16   The Upper Limb: From the Waist Up 177

    Bones and Joints of the Upper Limb 178

        Bones of the Shoulder Region *178*

        Joints of the Shoulder Region *181*

        Bones of the Arm and Forearm *182*

        The Elbow and Radioulnar Joints *183*

        Bones and Joints of the Wrist and Hand *184*

    Muscles of the Upper Limb 186

        Muscles of the Shoulder Region *186*

        Muscles of the Arm *188*

        Muscles of the Forearm *189*

        Muscles of the Hand *192*

        The Carpal Tunnel *194*

    Dem's Da Pits: The Axilla and Its Contents 194

        Blood Supply to the Upper Limb *194*

        The Nerve Supply to the Upper Limb *197*

17   The Pectoral Region and Thorax: The Chest Area 201

    On the Surface of Things: The Breast and Pectoral Region 202

        The Quadrants of the Breast *203*

        Lymphatic Drainage of the Breast *203*

    The Thoracic Wall 204

        Openings of the Thoracic Cavity *205*

        The Ribs and the Intercostal Spaces *205*

    Whatcha Got Under the Hood?: The Thoracic Organs 206

        The Heart *207*

        The Lungs and Pleura *209*

    "He Gasped": The Mechanics of Breathing 212

18   The Abdomen: The Belly of the Beast 215
      Scaling the Wall: The Anterior (Front) Abdominal Wall 215
          Planes and Quadrants: "It Hurts Here!" 216
          Muscles of the Anterior Abdominal Wall 217
      That Takes a Lot of Guts!: The Gastrointestinal (GI) Tract 218
          Esophagus: Food Tube 218
          Stomach: Acid Wash 220
          Spleen 220
          Small Intestine: Absorbing Nutrients 220
          Large Intestine: The Dryer 220
      Just There to Help: Accessory Organs of the Digestive System 221
          Liver: Metabolism and Detox 221
          Gallbladder: The Storage Tank 222
          Pancreas: The Juice Bar 223
      Up Against a Wall: Organs of the Posterior Abdominal Wall 224
          Kidneys and Ureters: Urine Production and Transport 224
          Adrenal Glands: Jolting You into Action 224

19   The Pelvis and Perineum: Stuck in the Middle of You 227
      A Bony Bowl: The Pelvis 227
          The Pelvic Openings 229
          You Floor Me: The Muscular Pelvic Diaphragm 229
      Lined Up in the Midline: The Pelvic Organs 229
          Ureters and Urinary Bladder 230
      Rectum 231
      Male Pelvic Organs 231
          Female Pelvic Organs 232
      A Diamond in the Rough: The Perineum 235
          The Urogenital (UG) Triangle 235
          The Anal Triangle 236

20   The Lower Limb: From the Waist Down 239
      Dem Bones, Dem Bones: Regions and Bones of the Lower Limb 239
          Butt Out!: The Hip and Gluteal Region 240
          Let's Thigh One On!: The Thigh 242
          Getting a Leg Up on the Competition: The Leg 245
          The Ankle and Foot: Carry That Weight 247
          I'm Hip to That: The Hip Joint 249
          Knee Deep: The Knee Joint 250
      Blood Supply to the Lower Limb 251
      Nerve Supply to the Lower Limb 253

21 "On Top of Old Shoulders": The Head and Neck.......................................257

A Head Case: The Skull and Its Openings...257

The Bone-to-Skin Connection: The Muscles of the Face and Scalp...........261

You Fill Up My Senses 262

I See a Vision: Orbit, Eyeball, and the Sense of Sight................................262

I Didn't Odor This!: Nose, Nasal Cavity, and Sense of Smell 266

This Bud's For You!: Mouth, Tongue, Salivary Glands, and Sense of Taste......268

I'm All Ears: The Ear and Sense of Hearing and Balance 272

The Neck: The Great Connector...274

Cervical Triangles 274

Cervical Viscera (Neck Organs) 275

Appendixes

A   Glossary...277

B   Further Reading...291

Index 295

Introduction

Of all the areas of scientific study, perhaps none is as complex, fascinating, and important as the study of human anatomy. An understanding of anatomy is essential preparation for many of the fastest-growing and best-paying health-care jobs in our economy. And as the baby boom generation ages, health-care workers will be more in demand than ever before in history.

But as I said, the human body is very complex, with a lot of intricate detail. However, those who just want to memorize lists of terms are not getting the full benefit of this knowledge. It's much more important to understand how the body parts and systems work together to form the incomparable, elegant, and amazing human machine.

This book lays out the foundations for understanding how the body is put together and how it works. It does so in a way that is as unintimidating and clear as possible. I start with the four basic tissues of the body from which our organs are made, progress to the major organ systems (systems anatomy), and end with studies of each functional region of the body, putting everything you've learned in context. And if at all possible, we'll have a little fun along the way!

Extras

Interspersed throughout the book and complementing the text, these sidebars provide pertinent and interesting information about a variety of clinically related anatomic topics. These will help reinforce learning and make the anatomy easier to remember.

Body Language

Anatomy is full of unwieldy terms and new concepts. These sidebars provide concise definitions of the most essential words you'll encounter in the book.

Foot Notes

These sidebars provide fascinating side notes, historical information, and extended details about the topics in each chapter.

Bodily Malfunctions

A complex machine like the human body is bound to experience breakdowns. These sidebars point out some of the more common diseases, conditions, and injuries that can occur in a particular part or system of the body.

Acknowledgments

This has been an amazing journey, but certainly not one that was traveled alone. First, I would like to pay special "thanks" to Tom Stevens, Acquisitions Editor, for providing me this opportunity to realize a personal dream, and to Julie Bess, Development Editor, and Jan Lynn, Senior Production Editor, for guiding the final disposition of this work. Thank you also to all the hardworking "behind the scenes" people at Alpha for your individual technical and creative expertise on my behalf.

I would like to recognize Lippincott Williams and Wilkins for their permission to use the wonderful LifeART images, and Sharon Teal and Christopher Brown, medical illustrators, from the Indiana University School of Medicine Office of Visual Media for their many excellent drawings and illustrations.

I would like to thank my many students over the years who have perhaps taught me as much as I've taught them, and to my school, Indiana University School of Medicine, for the opportunity to teach the subject I most love.

I owe a particular "mountain" of gratitude to Lori Cates Hand, my publishing and writing coach, who worked so hard and gave so much to this process. I can't imagine what I would have done without her encouragement, direction, and timely humor!

I want to also thank my wife, Linda, for her steadfast patience, encouragement, and gracious assistance over the past several months. And I am indebted to my friends at Bethel Lutheran Church for their prayers and support during this process. Finally, I thank God for the unique gifts and talents He's blessed me with. To Him be the glory.

Special Thanks to the Technical Reviewer

Special thanks are extended to my colleague, Ronald L. Shew, Ph.D., for his careful eye and constructive comments during his review of this book to ensure that it gives you everything you need to know about anatomy. Ron is an expert anatomist, with 25 years of experience teaching students in medicine, dentistry, and undergraduate allied health sciences. During his academic career, he has been the recipient of numerous teaching awards from peers and students at four major health science centers.

Trademarks

All terms mentioned in this book that are known to be or are suspected of being trademarks or service marks have been appropriately capitalized. Alpha Books and Penguin Group (USA) Inc. cannot attest to the accuracy of this information. Use of a term in this book should not be regarded as affecting the validity of any trademark or service mark.

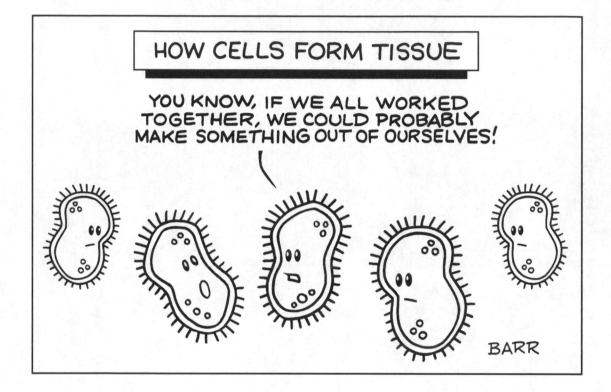

Part 1

Getting a Lay-of-the-Land View

Human anatomy is probably one of the most complex subjects you can study. So it helps to get your bearings first before you delve into the details.

This part starts off by giving you an overview of the subject of anatomy. Then you learn about the cells that make up the basic tissue types from which the body's organs are made. There are only four types of tissue in the body, and we've got everything you need to know about them right here.

In This Chapter

♦ The history of anatomy through the ages

♦ The different sub-disciplines and specialty areas of anatomy

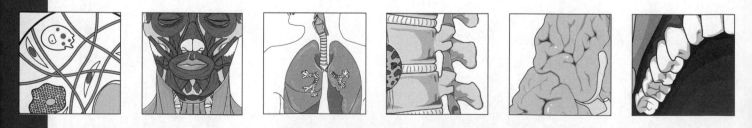

What Is Human Anatomy?

Knowledge about the human body underlies nearly every aspect of patient examination, diagnosis, and care. Despite its history as the oldest of the medical sciences, anatomy lives on as a vital, essential body of knowledge to be learned by a diverse group of health care practitioners. Studying anatomy is not easy. The marvelous complexity of the human frame has revealed itself rather grudgingly over the eons. New information continues to be added, while occasionally, old dogmas are challenged and revisited in light of new technologies.

A study of the human body is a study in a new and oftentimes foreign language. At its core, anatomy is a descriptive science. It's been estimated that the average medical student learns 10,000 new terms within his or her first years of medical school, many of which are rooted in anatomy. For health care specialists, knowledge of anatomy is an important tool of the profession as they provide competent care to patients.

Finally, the expression, "knowledge is power," as applied to becoming more informed about the human body, means that a person committed to learning the foundations of this subject will be better able to advocate for his or her own care or that of family and friends in today's health care arena.

A Brief History of an Old Subject

Evidence that early civilizations possessed knowledge of human structure has been largely deduced from Stone Age cave drawings depicting animals. Most of the early civilizations—such as the Babylonians, Egyptians, and Chinese—forbade desecration of the human body for religious or cultural reasons. So not many advances were made in the field during those times.

Early Greeks: Herophilus and Erasistratus

The first records of scientific discoveries that were meant to reveal information about the body and diseases related to it were attributed to the third century B.C.E. Greek philosopher-surgeons, Herophilus (c335–280 B.C.E.) and Erasistratus (c304–250 B.C.E.), who dissected bodies of condemned criminals. Most anatomic knowledge prior to that time was based on animal dissections. Herophilus is considered by many to be the father of anatomy. He and Erasistratus were founders of the medical school in Alexandria, Egypt. Herophilus is credited with identifying the brain as the seat of intelligence and with naming the duodenum. Erasistratus is believed to have discovered the heart valves and to have made important observations about nerves leaving the brain. He also believed that the arteries carried vital air, or pneuma, instead of the humors, which was the prevailing theory at that time. Their writings were unfortunately lost when the library at Alexandria was destroyed in 272 C.E.

However rational the minds of those early investigators might have been, they seemed no match for the irrational philosophies of the day that were steeped in the mystique of animal entrails and what they foretold for the next growing season. Anatomical study and dissection ground to a halt in the centuries following their work.

Physician to Gladiators and Dissector of Pigs: Galen

A resurgence in anatomy occurred in the second century C.E. in the person of Galen (c130–201 C.E.), who was chief physician to the Roman gladiators. History records that Galen's considerable studies were confined to dissections of pigs, apes, and other animals, not humans. This did not stop him from freely extrapolating what he found in animals and assuming that it accurately applied to the human. Although he correctly dispelled the notion that arteries carried air instead of blood, and observed that muscle contraction was caused by nerves, he incorrectly believed that blood flowed to and from the heart in an ebb-and-flow manner. He also described a *rete mirabile* (a close complex of arteries and veins) beneath the brain in humans, which was only present in the pigs he dissected.

Galen was a prolific writer and essentially rewrote the text on anatomical understanding that was to dominate thought for the next 1,300 years. His ideas were sacrosanct and above criticism. So powerful was his influence that once medical dissection of human cadavers was permitted in European medical schools in the thirteenth century, anatomists commonly discounted their observations at an autopsy if they conflicted with Galenic teachings. A real-life "don't confuse me with the facts" story.

A portrait of Andreas Vesalius.

(Photo courtesy of the National Library of Medicine)

The Modern Father of Anatomy: Vesalius

Finally, the true, modern day father of anatomy came along. His name was Andreas Vesalius (1514–1564). Vesalius was a Flemish physician-anatomist who was able to successfully challenge the centuries-old teachings of Galen. This was made possible by approval from the Catholic Church to dissect convicted criminals following death by hanging and by Vesalius' philosophy of performing these dissections himself and then accurately recording what he saw.

Vesalius also enlisted an artist to prepare drawings of his dissections and integrated these in a treatise he wrote. The product of these efforts was the first comprehensive textbook of anatomy called *De Humani Corporis Fabrica* (1543), which translates to *On the Fabric of the Human Body*. His critical approach to his studies set in place the importance of objectivity, which is central to the scientific method of inquiry. Despite such rational thought, Vesalius was a target of rebuke for some years before reason finally prevailed.

Under the Microscope: Leeuwenhoek

The invention of the microscope by Antonie Van Leeuwenhoek (1632–1723) ushered in a new way to view and study anatomy. His microscope allowed investigators to see cells and tissues with an "aided eye." This advance opened a whole new area of study, which became known as *microscopic anatomy* or *histology*. By taking small samples of tissues, anatomists could further explore the nooks and crannies of the body to learn how it was put together on a cell and tissue level. This has evolved and expanded considerably over the ensuing centuries so that today, anatomists, using sophisticated transmission and scanning electron microscopes can look at the fine details of surfaces of cells as well as their subcellular components.

Cover page of Vesalius' textbook, *De Humani Corporis Fabrica*, 1543.

(Photo courtesy of the National Library of Medicine)

The "ologies" of Human Anatomy

The word "anatomy" comes from the Greek word *anatome*, meaning literally, "a cutting up, or dissection." Anatomy is the study of the structure of living beings, which has come about over the eons of human existence by dissection of dead plants and animals, including humans. Human anatomy is the descriptive study of what we are made of!

Through the steady accumulation of knowledge about the structure and form of the human body that has been further advanced by new technologies and methodologies in other science and medical disciplines, anatomy has evolved into a broad, complex of subdisciplines or "ologies," where modern-day anatomists specialize. Underlying the investigations in each of these areas is how new observations and findings support or require modification of our understanding of structure. Here are a few of the most prevalent of these "ologies":

◆ **Gross anatomy.** The term "gross," as applied here, means "big," not "yucky." That is, it is a study of the structure and organization of the body without the aid of magnification (eyeglasses don't count!). It is macroscopic anatomy. A major component of this area of anatomy involves dissection of cadavers, and presently relies on the selfless and altruistic gift of human donors to advance medical research and teaching at medical centers and universities. This branch of anatomy was all that existed up until the time of Leeuwenhoek's invention of the microscope.

◆ **Histology.** The word part, "histo," refers to "tissue." This is the study of the tissues and cells of bodily fluids. Analysis of these requires a magnification aid of some kind. A similar term is *microscopic anatomy*. In order to improve viewing and interpretation of what is being examined, investigators have had to devise ways of cutting tissue into thin enough sections so that light can pass through the tissue section. This necessitated the development of embedding media, such as paraffin wax, or various resins that could hold the tissue so that it could be cut with a microtome, and stains to enhance the features of the examined tissue. Discoveries in the field have also come through advances in glass lenses and optical systems of microscopes.

◆ **Cell biology.** This is a cousin to histology. In this area of study, scientists extract or harvest cells from tissues for examination. Using a variety of microscopic, biochemical, molecular, or immunological techniques, they "dissect" the structure and function of subcellular components.

◆ **Embryology.** Embryology, or developmental anatomy, is the study of human development from conception to birth. In this area, scientists learn how organs and individual parts of the body develop, and when, and what biological signals control these processes. Present-day studies focus heavily on the genes responsible for controlling growth processes and how they are affected in health and disease.

◆ **Neuroanatomy.** Neuroanatomy involves study of the nervous system. The study of the brain and its complexity has been a recent frontier of anatomy and medicine. New advances are continually being made.

Astonishing, recent advances in radiological imaging, such as computed tomography (CAT scan), magnetic resonance imaging (MRI), positron emission tomography (PET scan), and ultrasound, now allow doctors to see with exquisite detail and clarity the anatomy of their patients. Such advances are major breakthroughs in patient care because they reduce the need for surgery and other invasive procedures to determine what is going on in a person's body.

Anatomy, like a fine wine, improves with age. Although it may be an old subject, it will never be a relic that is moved to the side because it has lost its relevance. To the contrary, medical, radiologic, and surgical advances will always be rooted in an understanding of anatomic structure and relationships. Anatomy, in all its manifestations, remains a vital contemporary science and is the cornerstone of medicine and its practice.

The Least You Need to Know

◆ The Greek physician Herophilus was the early "father of anatomy."

◆ Galen was the chief physician of the Roman gladiators and had an overwhelming, often negative, influence on anatomical thought in his day that stretched for 1,300 years.

◆ Vesalius was the person who refuted the teachings of Galen and moved the study of anatomy toward the contemporary scientific method approach.

◆ There are many subdisciplines in anatomy.

In This Chapter

◆ The types, location, and function of epithelial tissue

◆ The components and types of connective tissue

◆ The three kinds of muscle cells in the body

◆ The cells and features of nervous tissue

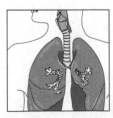
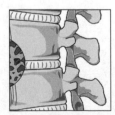
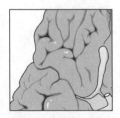

Chapter 2

Cut from the Same Cloth: The Four Types of Tissue

Four types of tissue—epithelial tissue, connective tissue, muscular tissue, and nervous tissue—are what everything in the human body is made of. Each of these tissues serves important but diverse functions. Amazing, isn't it?

Tissues are similar cells working together to support a particular purpose or function. This chapter highlights the distinctive features of each of these tissues.

Epithelial Tissue: The Great Shoreline of the Body

Imagine yourself in the great outdoors taking a boat ride on a large lake with many bays and inlets. In this metaphor, the "land" you are following along represents the body and the shoreline represents the epithelium. The shoreline covers the land (body) and lines all the bays and inlets you investigate on your trip. Entering any bay or inlet is equivalent to entering one of the many openings of the body (such as the mouth, nose, ear, anal canal, and so on). Even though the shoreline changes in its appearance during your trip—it may be a sandy beach, rocky, or tree-lined—you could never reach the body without first landing your boat on the epithelial shoreline.

The outer-surface epithelium is connected to the epithelial surfaces that lead into and line certain interior organs. The specific type of epithelial tissue changes according to the function it provides to the body. In its simplest function, epithelial tissue forms an unbroken boundary between the external and internal environments. It is also a barrier that germs must penetrate before they can attack the body. All body functions (like breathing, absorption of nutrients, and formation of urine) occur across epithelial surfaces. So, this great shoreline is very important indeed!

Epithelia on a Need-to-Know Basis

Here are the five most important things you should know about epithelial tissue:

1. Epithelial tissue covers and lines the external and internal surfaces of the body.

2. To be effective at this, epithelial cells need to be tightly connected to one another. A variety of specialized structures, called cell junctions, bind cells firmly to one another.

3. Because the cells are tightly connected, there is very little free space between adjacent cells. These last two features help the cells to act as a strong barrier.

4. Epithelial cells attach to a thin connective tissue sheet called the basement membrane. The basement membrane separates sheets of epithelial cells from other underlying tissue.

5. Epithelial cells are polarized, meaning that they have different surfaces and a particular internal organization of their organelles according to their function. Their bottom surface is attached to the basement membrane. Their side surfaces are in contact with adjacent cells, and their free surface is exposed to the external or internal environment.

The bottom line is that epithelia are specialized cell barriers that cover and line a variety of structures in the body and serve different functions, as you will learn next.

The Different Types of Epithelial Tissue

Epithelial cells differ in their size and appearance. They can be organized into a single row of cells or many rows of cells piled on top of one another. These differences are used to describe the various types of epithelial tissues in the body. These types are the following:

◆ **Simple epithelia.** A simple epithelium consists of a single row or layer of cells. The most common simple epithelium is simple squamous. This epithelium consists of very thin or flattened cells. Such epithelia line the internal surfaces of blood vessels and the body cavities. Because this epithelium is so thin, it also serves as an excellent lining for the air sacs in the lungs, allowing for the exchange of oxygen and carbon dioxide during breathing.

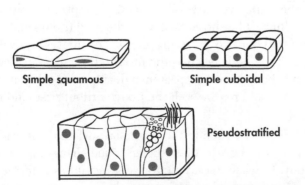

Simple squamous   Simple cuboidal   Simple columnar

Microvilli

Pseudostratified

Types of simple epithelia.

Body Language _____

Simple layers of cells that look box-like are called **simple cuboidal epithelia.** They commonly line portions of ducts of glands and tubules in the kidney. **Simple columnar epithelia** are so named because the cells are taller than they are wide. This kind of lining is most prevalent in many of the organs of the digestive system and in the female reproductive system.

◆ **Stratified epithelia.** This type of epithelia has several rows or layers of cells. The most prevalent type of stratified epithelium in the body is called stratified squamous epithelium. This type of epithelium covers the surface of the skin and the lining of the mouth, throat, esophagus, vagina, and anal canal. Such a lining occurs where tissues experience much wear and tear.

Stratified squamous epithelium.

Foot Notes _____

The stratified squamous epithelium of the skin differs from stratified squamous epithelium elsewhere in the body. Its cells gradually become filled with a water-resistant protein called keratin. This process is known as keratinization.

◆ **Pseudostratified epithelium** The passageways of parts of the nasal cavity, trachea, and bronchi are lined by pseudostratified columnar epithelium, also called *respiratory epithelium*. Although this epithelium looks like it's made up of several layers of cells, careful studies have shown that it actually consists of a single layer of cells with different shapes and sizes.

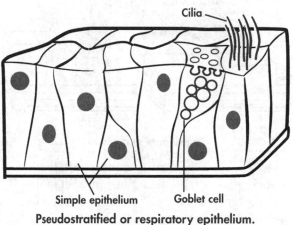

Cilia

Simple epithelium   Goblet cell

Pseudostratified or respiratory epithelium.

All the cells in this epithelium attach to the basement membrane. Several of the columnar cells have cilia, membrane-covered moving structures. Interposed among the cells are *goblet* cells. These cells have the appearance of a wine glass and are actually single-cell glands that secrete *mucin*. Mucin, when combined with water, produces the viscous fluid called mucus that helps trap inhaled particles and other debris. The wave-like action of the cilia sweeps trapped material out of these passages.

In addition to cilia, some cells have surface specializations of the cell membrane called *microvilli*. These are slender, finger-like projections on the surfaces of cells. They increase the cells' surface area to enhance absorption of nutrients during digestion or transport of fluids and ions during urine formation. They are abundant on the surfaces of cells lining the small intestine and on the tubules in the kidney.

◆ **Transitional epithelium.** The epithelial lining of the urinary passages and bladder is called transitional epithelium. This tissue must be able to expand to temporarily hold the urine produced by the kidneys. The outermost cells are shaped like cubes when the bladder is empty. They flatten as the bladder fills.

Transitional epithelium.

As you can see, there are many different types of epithelial tissue. The epithelium lining the "great shoreline" changes or specializes depending where it is located in the body.

The Tie That Binds: Connective Tissue

The connective tissues are the "glue" that provide the structural and nutritional support for cells, tissues, and organs of the body. Muscles are wrapped in connective tissue, epithelial surfaces rest on connective tissue, and nerves and blood vessels are distributed in the body within connective tissue. In addition, connective tissue plays an important role in the immunological defense of the body. Connective tissue is present throughout the body and is vitally important to its support and maintenance.

Although there are different kinds of connective tissue, they all share the following common features:

◆ Cells
◆ Fibers
◆ Ground substance

Connective Tissue Cells

Unlike muscle, nerve, or epithelial cells, connective tissue cells don't usually directly attach to one another. These cells have a lot of space between them. Within this space are varying amounts of *extracellular matrix*, which is made up of fibers and ground substance (more on these in a minute).

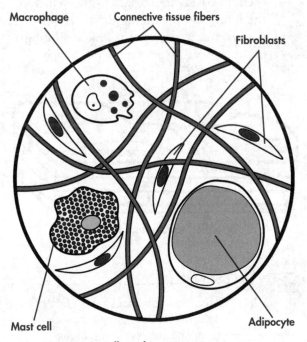

Common cells within connective tissues.

A variety of cells are found within connective tissue. Each has important and unique roles. Here's a rundown of these different cell types:

◆ **Fibroblasts.** The principal cell type found in connective tissue. They produce and secrete collagen, the most common protein in the body, as well as other non-collagenous proteins. Fibroblasts are very important for normal growth and development of the body and in response to tissue injury. Excess collagen from fibroblasts is what causes scars on the skin.

Foot Notes

Fibroblasts are cousins to skeletal cells called chondroblasts and osteoblasts. These connective tissue cells produce specific kinds of extracellular matrix called cartilage ("chondro") and bone ("osteo"), respectively. These cells are described in further detail in Chapter 4.

◆ **Adipocytes.** Adipocytes, also known as fat cells, are specialized connective tissue cells that store lipids (fatty acids). Their cytoplasm is mostly a single droplet of fat. Adipocytes are found throughout connective tissue spaces as a sort of natural packing material. When a bunch of them get together in one place, it's called *adipose tissue*. Adipose insulates the body against extreme cold temperatures and provides support around certain organs, like the kidneys and the eyeball.

◆ **Macrophages.** Macrophages (Greek for "mighty eaters") are another cell type common to connective tissue. These cells have great appetites for cleaning up dead or damaged cells, bacteria, or foreign substances that enter the body. They even eat cancer cells. These cells also produce *cytokines*, which are proteins that can enhance

other cells in the immune system. They get into connective tissue spaces from the bloodstream through the walls of capillaries.

◆ **Mast cells.** Mast cells are most numerous in connective tissue spaces of the skin, conjunctiva (covering) of the eye, the nasal cavity, and lining of the lungs and digestive tract. They contain many granules that are loaded with histamine and other potent chemicals. Mast cells play an important role in allergic reactions and in the inflammatory process. When they are stimulated, they quickly release their chemicals, which cause local tissue swelling, redness, and itching.

◆ **Leukocytes.** Leukocytes (or white blood cells) aren't normally in connective tissues, but readily migrate from vessel walls into connective tissue spaces in response to injury or inflammation.

Connective Tissue Fibers

Three main kinds of fibers are produced by fibroblasts within connective tissue:

◆ **Collagen.** Collagen is the most abundant protein in the body. It is found in high concentrations in structures that require tensile strength and rigidity, such as bones, teeth, tendons, ligaments, and the dermis of the skin.

◆ **Reticular fibers.** Reticular fibers consist of collagen and are much thinner than regular collagen fibers. They form a mesh-like framework for *hemopoietic* and lymphoid organs, such as bone marrow, spleen, and lymph nodes.

◆ **Elastin fibers.** Elastic fibers, produced by the protein elastin, are the rubber bands of connective tissue. They are found in high concentrations in tissues or structures that stretch and recoil, such as the wall of the aorta, the dermis of the skin, the epiglottis of the larynx, and the cartilages of the ear.

These fiber types are unevenly distributed. Connective tissues are classified according to the density (loose or dense), arrangement (regular or irregular), or predominance of fiber type (collagenous, elastic).

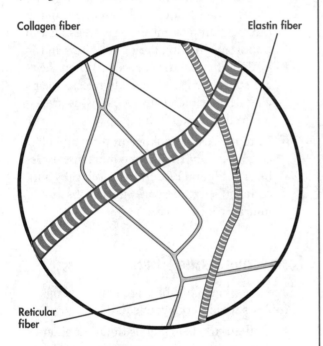

Three main kinds of fibers in connective tissue.

Ground Substance

Ground substance is a colorless, gel-like substance that consists mainly of water and non-collagenous proteins. The cells and fibers of connective tissue are embedded in it. Ground substance functions as an important barrier to invading microorganisms. It also facilitates normal exchange of nutrients and metabolic waste products between cells and tissues.

Types of Connective Tissue

Types of connective tissue are identified based on the relative amounts, arrangement, and type of fibers present.

Ordinary or connective tissue proper consists of two types: loose and dense.

Loose connective tissue is found throughout the body. It is best seen in the hypodermis or subcutaneous tissue layer between the skin and muscles. This tissue contains variable amounts of fat cells within a delicate meshwork of fibers. It provides support to epithelial tissue and encloses blood vessels and lymphatic channels.

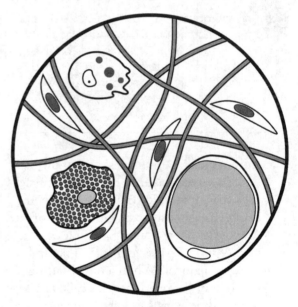

Loose connective tissue.

Dense connective tissue differs from loose connective tissue by having fewer cells and more abundant thick bundles of collagen fibers. Such tissue is strong and well suited to resist stress and forces applied to it.

Dense connective tissue is identified as regular or irregular, depending on how the fibers in it are arranged. In dense regular connective tissue, collagen fibers are arranged in a parallel, linear fashion, such as in tendons and ligaments.

Collagen fibers   Fibroblasts

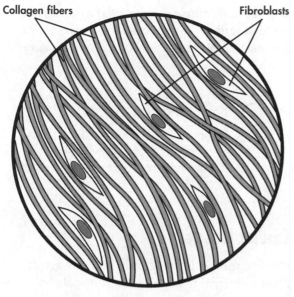

Dense regular connective tissue.

Dense irregular connective tissue is found in the dermis of the skin. Here, collagen fibers are randomly arranged, which gives this tissue its ability to resist pulls from all directions.

Collagen fibers   Fibroblast

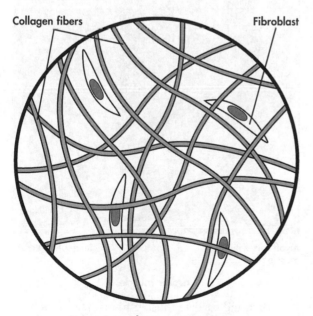

Dense irregular connective tissue.

Foot Notes

The special connective tissues of cartilage and bone are discussed in greater detail in Chapter 4. And believe it or not, blood is also considered a connective tissue.

Pump Me Up: Muscular Tissue

People's lives depend on the action of muscles. The air they breathe, the rhythmic beat of their hearts, and the transport and digestion of the food they eat all rely on muscle action.

The structural unit of a muscle is a single muscle cell, also known as a *muscle fiber*. Muscle cells are full of proteins that make them contract.

Body Language

A **muscle fiber** is a single cell of a muscle. A single muscle fiber can be nearly a foot long.

There are three kinds of muscle in the body. Each differs in its location, appearance under the microscope, and function:

◆ Skeletal muscle

◆ Cardiac muscle

◆ Smooth muscle

The following sections give more details on each type of muscle.

Skeletal Muscle

If the bones of the body represent the means by which people get someplace quickly and efficiently, then it is the muscles attached to them that make it happen. Without them, a person would be nothing but a bag of bones—filled with good intentions, but no way to go.

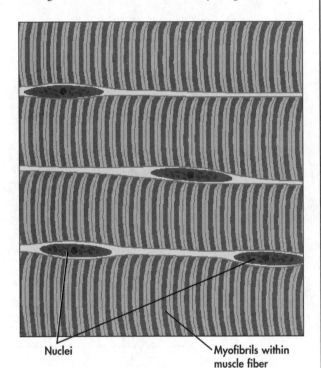

Skeletal muscle.

The majority of the muscles in the body are skeletal muscle. Skeletal muscles typically attach at each of their ends to bones, although some attach to the eyeball, the skin of the face and head, or the mucous membrane of the tongue.

Skeletal muscle cells are long and cylindrical. They contain many nuclei, which are located on the outside edges of the cell. Each cell contains an abundant, orderly arrangement of contractile proteins. When viewed with a microscope, this arrangement produces a pattern of alternating dark and light lines. This gives the tissue a characteristic striped appearance and explains

why skeletal muscle is also called *striated muscle*. Skeletal muscle is often called *voluntary muscle* because the person largely controls its contraction.

Body Language _____

Voluntary muscles are those that are consciously controlled by the person (for example, the skeletal muscles that move the arm).

Cardiac Muscle

Cardiac muscle is found only in the heart. It is the principal tissue forming the walls of the heart's chambers. It is best seen in the right and left ventricles that must pump blood to the lungs and body.

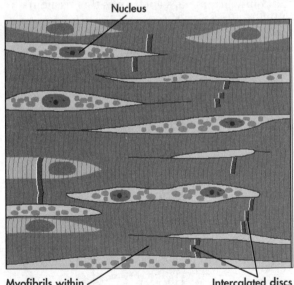

Cardiac muscle.

Individual cardiac muscle fibers are branched. They connect end-to-end with adjacent fibers by specialized cell junctions called *intercalated discs*. This arrangement is important for the coordinated contraction of the heart muscle.

Like skeletal muscle, cardiac muscle contains abundant contractile proteins and appears striated. However, unlike skeletal muscle, a single nucleus is located at the center of the cell. Cardiac muscle is further unique in that it is "born to beat" all on its own, without the need for nerves. Only the rate and strength of heartbeat is modified by the autonomic nervous system (or by cardiac drugs). Cardiac muscle is *involuntary muscle*, meaning that its contractions are not under a person's conscious control.

Body Language

Involuntary muscles are those that are controlled by the autonomic nervous system (such as heart muscle).

Smooth Muscle

Smooth muscle is found in the walls of the hollow organs (for example, the digestive, urinary, and reproductive systems) and the blood vessels. Smooth muscle is also found in the iris of the eyeball, where it controls the diameter of the pupil, as well as in the skin, where it attaches to hair follicles to produce goose bumps in response to cold temperatures or fear.

Muscle cells of this type are small and spindle-shaped. They lack the concentration of contractile proteins found in skeletal and cardiac muscle cells (and, hence, don't have striations), and contain a single, centrally placed nucleus. Contraction of these cells is under involuntary control by the autonomic nervous system.

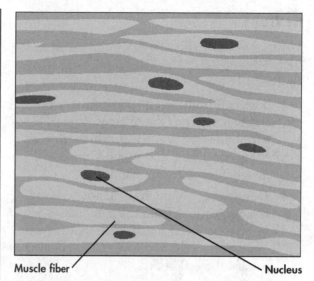

Muscle fiber   Nucleus

Smooth muscle.

Nervous in the Service: Nervous Tissue

If it is the muscles attached to bones that move the body in space, it is the nervous tissue that supplies the necessary electrical current to activate the muscles into action. Nervous tissue is made up of two principal cell types:

◆ Neurons (nerve cells), which form the structural and functional unit of the nervous system

◆ Neuroglial (or simply *glial*) cells, which support the neurons

Neurons

Most neurons consist of a cell body and specialized cell processes. There are three parts to a neuron:

♦ **The cell body.**   The part of the nerve cell that contains the nucleus. It is also surrounded by cytoplasm, which contains many organelles that support the cell's metabolic and functional needs.

♦ **Dendrites.**   Numerous slender cell surface projections from the cell body. They receive electrical signals or impulses from other nerve cells and relay these impulses toward the cell body.

♦ **Axon.**   A single process that carries impulses away from the cell body. An axon is a long process (some axons may be up to a few feet in length in tall people!) that ends by dividing into several terminal branches. These make special contacts, called *synapses*, with other nerve cells, muscles, or glands to produce a particular action.

Body Language

Synapses are the spaces between a nerve cell and another cell, through which impulses are transmitted.

Foot Notes

Most axons are enclosed by segments or intervals of a fat-rich insulating material known as *myelin*. Myelin helps to increase the speed of transmission of nerve impulses along axons. Axons with this covering are referred to as *myelinated axons*. Axons without this material are *unmyelinated axons*, and they send impulses at a slower speed.

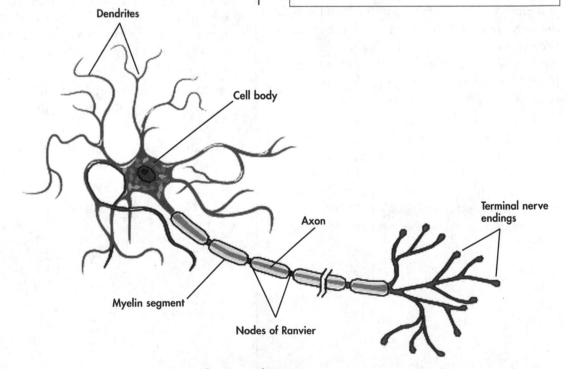

The parts of a neuron.

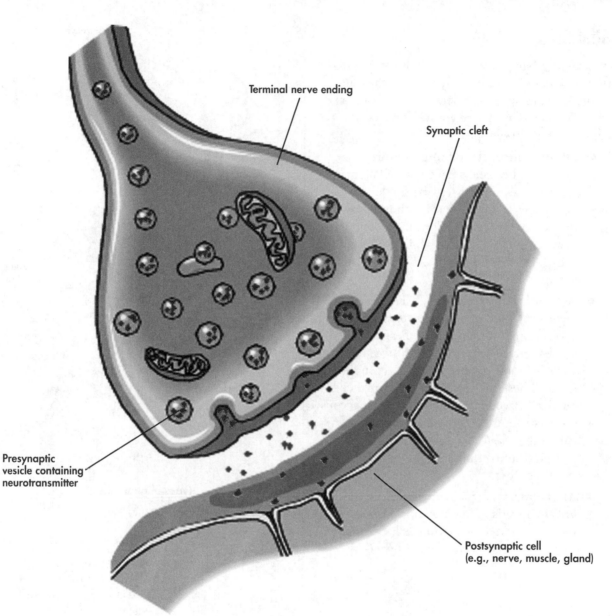

Terminal nerve ending

Synaptic cleft

Presynaptic
vesicle containing
neurotransmitter

Postsynaptic cell
(e.g., nerve, muscle, gland)

The synapse.

It is at synapses that the impulse signal is delivered to another cell. This *neurotransmission* is typically chemical. Special chemicals, called *neurotransmitters*, are stored in membrane sacs at the ends of axons. These substances are then released into a very narrow space, called the *synaptic cleft*, between the neuron and another nerve cell, gland, or muscle cell. These chemicals cross this space and are taken up by the recipient cell. The chemical causes a change in the membrane properties of the recipient cell, which either continues the signal or stops it at that point.

Glial Cells

Neurons are supported by an abundant population of *glial cells* that literally serve as the glue of the nervous system. There are ten times as many glial cells are there are neurons. Glial cells come in three types:

◆ **Oligodendroglia, or oligodendrocytes.** These cells have few cell branches or processes. Processes of a single oligodendrocyte wrap themselves numerous times around several axons in the central nervous system (CNS), while Schwann cells form myelin around only a single axon.

◆ **Astroglia, or astrocytes.** These cells have a star-like pattern of cytoplasmic projections. At the ends of these processes are *foot processes* that attach to structures, such as blood vessels, that run through the CNS. These processes form a virtual cell barrier between the CNS and non-nervous tissue structures. This produces what is called the *blood-brain barrier*. This barrier prevents contact between blood and brain tissue spaces.

◆ **Microglia.** As the name suggests, these are small cells. They function like macrophages and attack dying cells or microorganisms that have invaded the CNS.

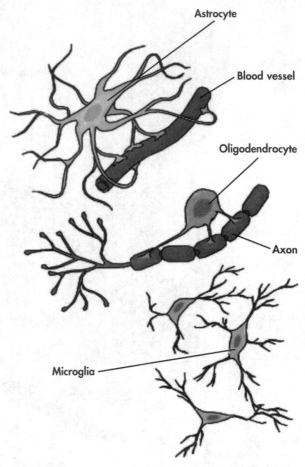

Types of glial cells.

Bodily Malfunctions

Multiple sclerosis is a neurological disorder where there is patchy loss of the insulating myelin around neurons in the central nervous system. This leads to interruptions in the transmission of nerve signals and to loss of muscle control.

The Least You Need to Know

◆ The organs and systems of the body are made from four types of tissue: epithelial tissue, connective tissue, muscle tissue, and nervous tissue.

◆ Epithelial tissues are either simple or stratified layers of cells and serve a variety of functions according to their type (for example, protection, diffusion, transport, absorption, or excretion).

◆ Connective tissues consist of cells and fibers contained within a ground substance and are named according to the principal kinds of cells, the kinds and density of fibers, and the direction of the fibers.

◆ There are three kinds of muscle tissue in the body, each of which is unique in microscopic appearance, specific function, and nerve supply.

◆ Nervous tissue consists of neurons—which relay impulse signals to other nerve cells, muscle cells, or glands—and supportive cells.

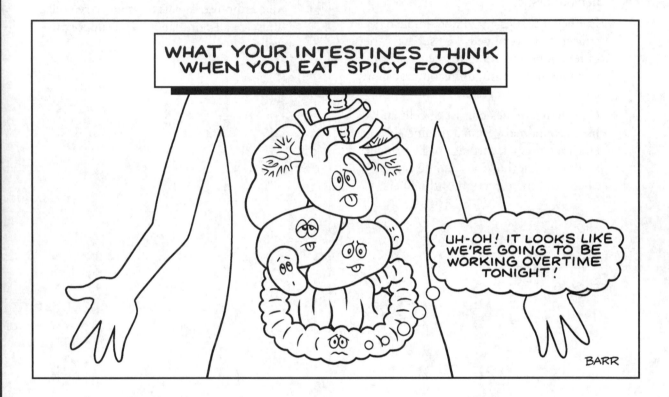

Part 2

All Systems Go!: Organs and Body Systems

When you think about learning anatomy, you probably think about the various systems, or groups of organs, that make up the body. There's the respiratory system that supplies oxygen to the body, the digestive system that creates energy, the skeletal system that holds it all together, and the muscular system that helps it move. And that's not even the half of it.

This part gives you a basic foundation for understanding the body's systems and their various parts. And there is a lot to know here.

In This Chapter

◆ The layers of the skin

◆ What skin does

◆ How skin makes hair, nails, and glands

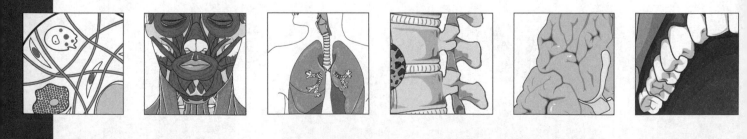

Only Skin Deep: The Integumentary System

The integumentary system (after the Greek word meaning "covering") consists of the skin, the hair, and the nails. Although you don't automatically think of the skin as an organ, in fact, it is. It is made up of tissues that work together to perform specific functions. By weight, it is the largest organ of the body: it makes up about 16 percent of the body's total weight.

Well, if it's an organ, you might ask, what does it do? Several important things, actually. It is an effective waterproof covering that not only keeps people dry, but also prevents them from losing their body fluids and drying out. It protects the body from external injury, invading microorganisms, and the harmful effects of solar radiation. It also helps regulate the body's temperature. And because of its rich nerve supply, it acts as an extensive sensory organ, maintaining communication with the world around it. Finally, the important vitamin D, which is essential for building strong bones and teeth, is produced in the skin following exposure to ultraviolet light.

In this chapter, you will learn about the layers of the skin and how they works. You will also learn about some of the other things (glands, hair, and nails) that are formed by skin.

The Skinny on Skin

The skin is made up of two layers: the *epidermis* and the *dermis*. The epidermis is the outermost epithelial layer. It consists of many layers of cells. The outer layers of cells are filled with keratin and are scale-like. The epidermis attaches to the dermis layer, which is rather thick and tough. In fact, this is the "hide" of skin as you know it. The dermis is bound to an underlying cushion of subcutaneous tissue known as the *hypodermis*.

Body Language

The **hypodermis** connects the skin to the underlying muscles of the body. It is made of loose connective tissue and variable numbers of fat cells.

Epidermis: Where the Beauty Is

Just as beauty is sometimes described as being only skin deep, the epidermis is the thinnest layer of the skin. If you viewed the epidermis with a microscope, you would see many rows of cells piled on top of one another and grouped into five *strata* (or layers). These strata, starting from the top and working down, are the following:

◆ **Stratum corneum.** Consists of 15 to 20 layers of thin scales, which are continuously shed.

◆ **Stratum lucidum.** Cells in this layer are very flat and translucent. Here cells die and lose their nuclei and organelles when they become full of keratin.

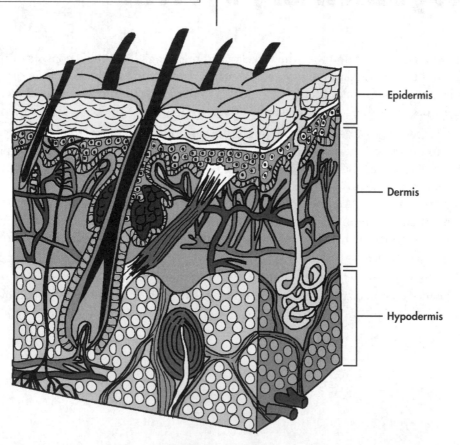

A view of the skin.

◆ **Stratum granulosum (granular cell layer).** Cells in this layer become filled with granules and secrete a waterproof lipid.

◆ **Stratum spinosum (prickle cell layer).** Cells in this layer are tightly joined by spinelike projections, which give them a prickly appearance. The cells contain filaments of keratin.

◆ **Stratum basale (germinative layer).** This is the deepest layer, resting on the basement membrane. It contains stem cells that divide and provide continual renewal of cells.

Most of the cells that make up the epidermis are called *keratinocytes* because during the growth, development, and life of these skin cells, they produce and accumulate within their cytoplasm a special resistant protein known as *keratin*.

Body Language

Keratin is a tough protein that makes skin strong. Continuous production and accumulation of keratin within a skin cell eventually kills it, leaving a cell remnant that contains keratin (a process called keratinization).

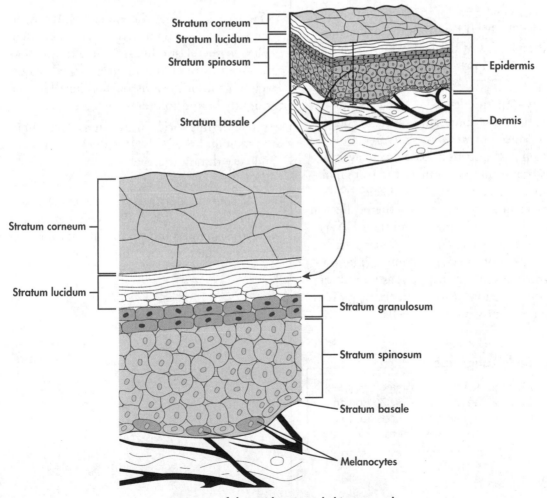

Layers of the epidermis and skin renewal.

Because the skin is subject to abrasion and the wear and tear of life, it must be capable of rapid renewal. The following figure outlines the process of skin cell renewal.

Bodily Malfunctions

Psoriasis is a common disease of the skin where there is uncontrolled multiplication of cells in the lower strata and a shorter time span between cell birth and death. This results in increased thickness of the epidermis. People with psoriasis have extensive red, scaly patches on their skin that are itchy and painful.

The epidermis varies in thickness depending on body region. It's thickest on the palms of the hands and soles of the feet, which undergo much abrasion and pressure during daily activities.

A variety of pigments are in the skin, but *melanin* is the most abundant kind. This pigment is produced by *melanocytes*, which are cells located in the stratum basale cell layer. Granules of melanin are transferred from melanocytes into the skin cells of the deepest rows of the epidermis. These granules then collect in the cells so as to shield them (and their DNA) from the harmful effects of ultraviolet radiation. Darkening of the skin ("tanning") following exposure to sunlight happens as a result of increased production of these melanin granules and their accumulation in skin cells.

Body Language

Melanin is the most common pigment found in the skin. It serves to darken the skin following sun exposure, to protect it from ultraviolet radiation.

The epidermis lacks blood and lymph vessels and gets its nutrition from vessels in the adjacent portion of the dermis.

Bodily Malfunctions

Albinism is a hereditary disorder that causes people to have no pigment in their skin, eyes, and hair. People with albinism are unable to produce melanin within their melanocytes. This puts them at risk for severe sunburn if their skin is not covered at all times.

The Dermis: What Lies Beneath

The dermis is under the epidermis. It's made of dense, irregular connective tissue that contains collagen and elastic fibers. These fibers allow the skin to stretch during pregnancy, significant weight gain, or local tissue swelling. The dermis is tightly bound to the overlying epidermis.

The dermis varies in thickness depending on the region of the body. It is thickest on the back and very thin in the eyelids.

The dermis has a rich nerve and blood supply. It contains derivatives of the epidermis, namely, hair follicles, nails, sweat glands, *sebaceous* (oil) glands, and smooth muscle structures known as the arrector pili muscles.

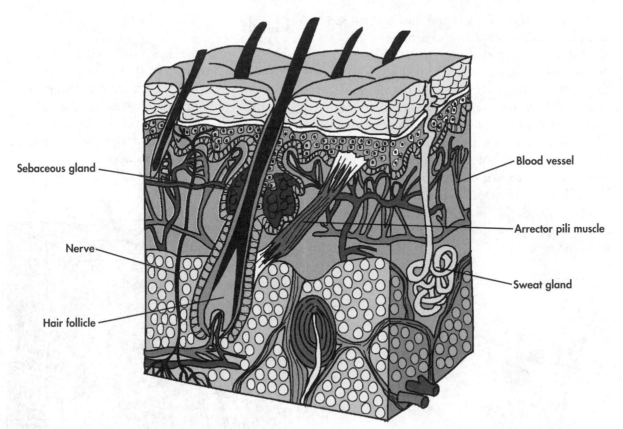

Sebaceous gland

Blood vessel

Nerve

Arrector pili muscle

Sweat gland

Hair follicle

The dermis.

Things Made from the Epidermis

Hair follicles, nails, sweat glands, and sebaceous glands are specialized structures of the epidermis that grow down into the underlying dermis layer but retain contact with the epidermis. The following sections discuss these structures in detail.

Hairs and Nails

Hairs are produced in *hair follicles* located deep within the dermis. Hair grows by cell division, which occurs in the *hair bulb* at the base of a follicle. This process of cell multiplication from below, and pushing older cells upward toward the epidermal surface, is just like the process for skin growth.

Body Language

Hair follicles are a part of the skin that grows hair by packing dead skin cells together and pushing them out toward the surface.

Beneath the hair bulb is a bud-like projection of connective tissue that contains tiny blood vessels. It's called the *dermal papilla*. The dermal papilla provides nutrition to the dividing cells in the hair bulb. In addition, melanocytes located in the hair bulb produce melanin, which gives hair its color. Melanin production slows and eventually stops as people age. That's why older people's hair appears white (so don't let your parents tell you it's *your* fault).

The ends of the fingers and toes are covered and protected by plate-like structures called nails. These firmly rest on a thin layer of epidermis called the *nail bed*. The nail bed is so thin that the blood vessels in the dermis produce the characteristic pinkish coloration of this portion of the nail. The root or base of the nail, where growth occurs, lies in the *nail groove* and is covered by a fold of skin. The free edge of this skin is the *cuticle*. At the bottom of the nail is a half-moon-shaped, whitish area called the *lunula*. Beneath the lunula are epithelial cells that multiply, grow, and become keratinized to form the nail.

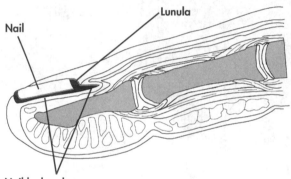

The parts of a nail.

Foot Notes

Associated with nearly all hair follicles in the skin are tiny bundles of smooth muscles called arrector pili muscles. As their name suggests, they erect the hairs to produce "goose bumps" when a person is frightened or shivering cold. These muscles receive their nerve supply from the "fight or flight" portion of the autonomic nervous system.

Glands

The surface of the skin is softened and protected by a thin film of oil produced by the sebaceous glands. Excess body heat is regulated, in part, by evaporation of sweat produced by sweat glands. The parts of them that control secretions, located in the dermis, are connected to the epidermis by their ducts.

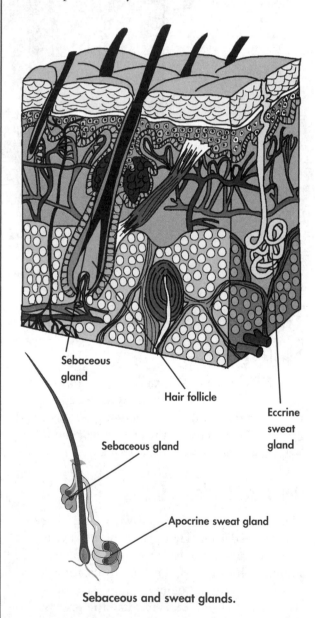

Sebaceous and sweat glands.

Following are the different types of glands located in the skin:

◆ **Sebaceous glands.**   Sebaceous glands are located throughout the body, but not in the skin of the palms of the hands or soles of the feet. Sebaceous glands are simple sac-like glands with a duct that opens to the upper end of a hair follicle. Cells in these glands produce an oily substance before they eventually die and disintegrate. This material, known as *sebum*, is then discharged through its duct.

Bodily Malfunctions

Acne is a common skin condition of adolescents that happens when blockage of sebaceous gland ducts causes the release of sebum into surrounding tissue. This results in inflammation and skin changes.

◆ **Apocrine sweat glands.**   These glands are limited in their distribution—they are located in the skin of the armpit, the pubic region, and the pigmented area of the breasts. After puberty, apocrine glands begin secreting a slightly viscous fluid that develops an odor following contact with bacteria on the surface of the body (hence, the rationale for using antibacterial deodorants).

◆ **Eccrine sweat glands.**   These glands are widely distributed in the skin. They produce a clear, watery secretion that contains dissolved salts and other substances. Eccrine sweat glands are particularly abundant on the palms and soles, explaining the clichés "sweaty palms" and "sweaty feet." These glands play an important role in regulating body temperature.

Two other types of sweat glands are the ceruminous glands, which produce ear wax, and the mammary glands, which produce milk. (See Chapter 13 for more on the latter).

The Least You Need to Know

◆ The skin is the largest organ of the body and has multiple functions.

◆ Skin consists of two layers: the epidermis and the dermis.

◆ Hairs and nails, and sweat glands and sebaceous glands are formed in the epidermal layer of skin.

In This Chapter

◆ The basics about bones and how they grow

◆ The two different parts of the skeletal system

◆ The two types of joints that hold bones together

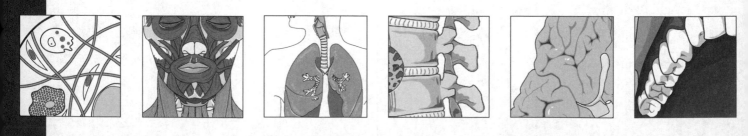

Chapter 4

Skeleton Crew: The Skeletal System

The skeletal system is the internal framework to which muscles and skin are attached—in other words, bones. It provides structural support and is powered by skeletal muscles, which enable the body to move. You can imagine that without this system of mechanical levers, movement would be severely restricted. The skeletal system is a multifunctional system. It protects vital internal organs, like the brain, heart, and lungs. It's the place where new blood cells are formed. And it's a rich reservoir of important minerals, particularly calcium and phosphorus, which are essential to the body.

The skeletal system is held together by various types of joints, which enable a person to move. Because the skeletal system is one of the most commonly injured and diseased systems of the body, it is important to understand how the skeleton is put together and how it works. This chapter does just that!

Bone Basics

When you think of bones, you think of something solid, like rigid columns of cement. Quite the contrary. Bones are actually specialized connective tissue that is always changing and growing. Bones are made up of living cells embedded within or lining the surfaces of a framework of minerals. They grow as the body grows and develops. And when they are broken (fractured), they are able to heal.

This section discusses the different types of bone cells, how bones grow and develop, and some common problems related to bones.

Call Me on My Cell

Like any living tissue, bone tissue contains cells. There are three different types of bone cells, each with its own function. They are:

- **Osteoblasts.** Cells that make new bone

- **Osteocytes.** Cells that are part of the bone

- **Osteoclasts.** Cells that break down and reabsorb old bone

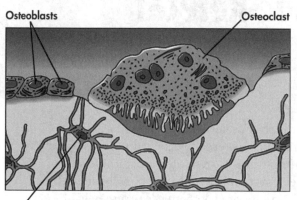

Osteoblasts

Osteoclast

Osteocyte

The three types of bone cells.

Osteoblasts (which is Greek for "bone germs") are responsible for bone formation. They produce the structural material that makes up bone. Osteoblasts stick to external and internal bone surfaces, where they actively produce an organic matrix (or framework) called *osteoid*. Osteoid is about 90 percent type-I collagen (the other 10 percent is made up of proteins and factors that regulate bone growth). It lacks bone minerals. Newly secreted osteoid quickly hardens, or mineralizes, as a result of the calcium and phosphates that are deposited on it. Thus, bone matrix is a composite material, which is why it is so strong and resilient.

Osteocytes (which is literally Greek for "bone cell") are former osteoblasts that became buried within the mineralized matrix they produced. Osteocytes are enclosed in tiny cavities called *osteocytic lacunae*. These cells maintain contact with neighboring osteocytes and osteoblasts on the surface of bone through slender extensions of the cell that are enclosed by tiny canals of bone called *canaliculi*.

Foot Notes

Over 90 percent of bone cells are osteocytes. Current research suggests that they play an important role in helping bones sense and adapt to changes in mechanical loads. They also are important for regulating mineral exchange between the bone and the blood.

Osteoclasts (Greek for "bone breakers") are large cells with 4 to 20 nuclei that specialize in reabsorbing or degrading bone matrix. In other words, they chew up bone. They break down old or damaged bone, allowing for new bone to be formed. In this process, they release calcium from the bone into the bloodstream to meet the body's requirement for this mineral.

Bone Growth and Development

Bone develops and grows by one of two processes:

- Intramembranous ossification
- Endochondral ossification

In *intramembranous ossification*, bone develops before birth within an embryonic connective tissue membrane. The embryonic cells that form this membrane become osteoblasts. The osteoblasts secrete bone matrix to form islands of bone that expand in a radial direction as more calcium and phosphates are deposited.

This type of bone development forms the flat bones of the skull, the bones of the face, and the clavicle or collar bone. After birth, it causes bones to grow in diameter.

Most bones, particularly the long bones of the limbs, are formed by a process called *endochondral ossification*. In this process, embryonic cells become *chondroblasts*, which secrete cartilage matrix. This matrix forms itself into a model of the future bone. Subsequently, the shaft region (*diaphysis*) of the future bone is mineralized, and blood vessels invade it. Here, the cartilage matrix is broken down and replaced with bone. This is the primary place where *ossification* (turning to bone) happens.

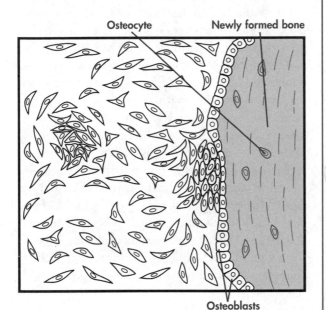

Intramembranous ossification.

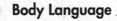

Body Language

Chondroblasts are cells that produce cartilage matrix (framework) and create a model of the bone, which is later mineralized.

The **diaphysis** is the shaft area of a long bone. Epiphyses are the ends of the bone.

Ossification is the process of replacing soft cartilage with bone matrix that mineralizes, thus forming real bone.

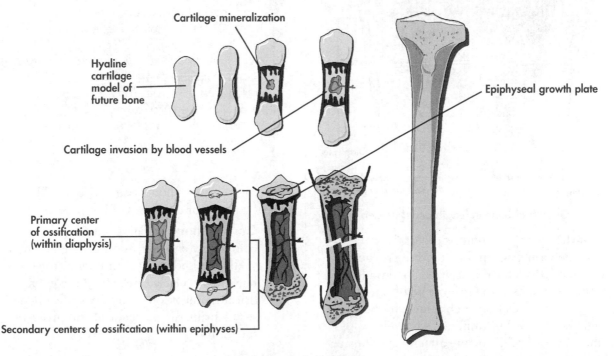

Endochondral ossification.

Shortly thereafter, similar events occur at the ends (epiphyses) of the cartilage model to form secondary centers of ossification. At these centers, two cartilage structures will remain after they have been replaced by bone—one temporary (the growth plate cartilage), the other permanent (the articular cartilage). The growth plate remains between the shaft and the ends for several years and enables the bones to grow in length.

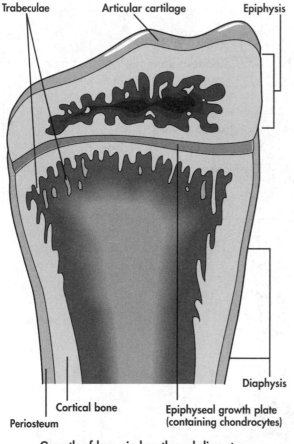

Growth of bone in length and diameter.

At the epiphyseal (end) sides of the growth plate, new cartilage and matrix are produced. This model is later mineralized and invaded at the shaft side of the growth plate. These invasive activities leave behind narrow beams (called *trabeculae*) of mineralized cartilage. These are scaffolds for the attachment of osteoblasts and the deposit of new bone on their

surfaces. These so-called *primary trabeculae* of mineralized cartilage and bone are progressively replaced by secondary trabeculae of bone that the osteoblasts produce.

At the end of adolescence or in early adulthood, the chondrocytes stop being created in growth plates. The plates ultimately disappear, so the ends and the shaft are now completely connected by bone. At this point, the bone can no longer grow in length. The diameters of bones increase via the addition of bone matrix by osteoblasts located on the external surface of bone within a connective tissue covering called *periosteum*. The ends of bones formed by endochondral ossification remain covered by cartilage. This is the pearly gray "gristle," like what you'd see at the end of a chicken bone. This is the material that bones contact and slide over at the joints in the body (more on this later in this chapter when I discuss joints).

Foot Notes

Bones come in different shapes and sizes. *Flat bones* form the roof of the skull, the scapula (shoulder blade), and the ilium (part of the hip bone). *Short bones* are found in the wrist (carpal bones) and foot (tarsal bones). *Irregular bones* include the vertebrae. *Long bones* are found in arms and legs (for example, the humerus, the radius, the ulna, the femur, and the tibia).

If you cut into a bone lengthwise, you'll see that it consists of two types of bone tissue:

◆ **Cortical bone.**   This bone is dense, completely encasing and forming an outer shell on all bones. It is thinner at the ends, but is particularly thick along the cylindrical shafts of long bones. Cortical bone serves primarily mechanical and protective functions.

◆ **Trabecular or cancellous bone.**   This bone appears as a spongy meshwork of tissue. It is concentrated at the ends of long bones or within irregular bones, such as the vertebrae, to give the bone additional support and strength. Bone marrow tissue fills these meshwork spaces, as well as the shaft. Blood cells are formed within this tissue.

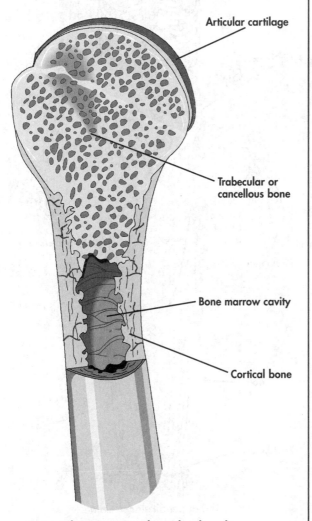

Articular cartilage

Trabecular or cancellous bone

Bone marrow cavity

Cortical bone

Lengthwise section through a long bone.

The Two Divisions of the Skeletal System

The skeletal system in an adult consists of about 206 bones. Anatomists have divided the skeleton into two systems in order to study it better: the *axial system* and the *appendicular system.* "Huh?" you might ask. Don't panic. Just consider the definitions of the terms *axial* and *appendicular,* and they will help you categorize which bones fall into which division.

The Axial Skeleton: Skull, Spine, Ribs, and Breastbone

The term *axial* refers to an axis—the center of something. In this instance, it refers to the central part of the body; that is, the head and trunk. Hence, the axial skeleton is made up of the bones in four structures:

◆ The skull

◆ The vertebral column (or spine)

◆ The ribs

◆ The sternum (to which the ribs attach—also called the breastbone)

Pretty straightforward, right? Yes, there are 28 bones in the skull and 33 vertebrae in the spinal column, but we'll worry about those details later, in Chapter 15. The most important thing to know about the axial skeleton is that it protects such vital organs as the brain, the spinal cord, and the heart and lungs. Got that?

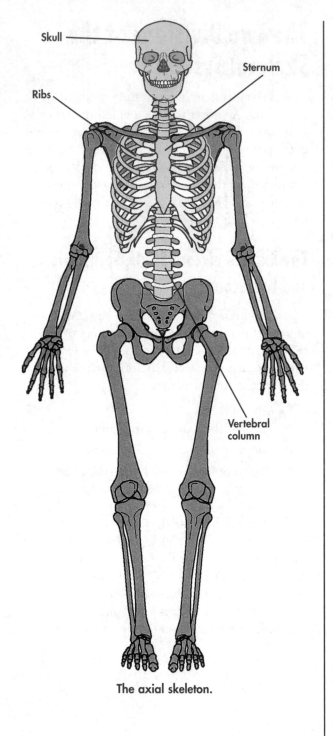

The axial skeleton.

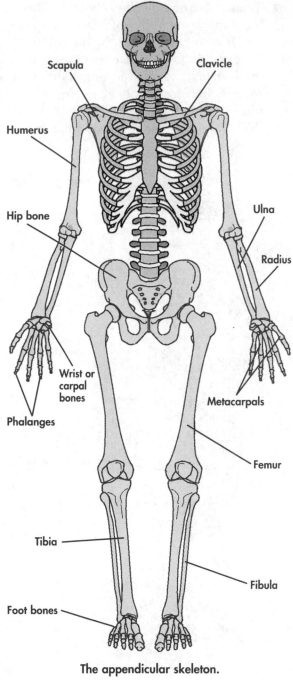

The appendicular skeleton.

The Appendicular Skeleton: Everything Else

The term *appendicular* relates to appendages; something that attaches to a main structure, such as the trunk of the body. So, bones that compose the appendicular skeleton are those associated with the limbs. What bones are we talking about? Well, basically, everything else that isn't part of the axial skeleton.

One easy way to remember which system a bone belongs to is to use the "rule-out principle." If you learn which bones are in the axial system, which is a shorter list, then you'll know that all the other bones have to be components of the appendicular system!

In Part 3, I discuss individual bones and particular features of bones as they relate to each body region.

Jumpin' Joints: Where Two Bones Come Together

Joints, also called *articulations*, occur at junctions between two or more bones. Joints help to do one or more of the following:

- ◆ Facilitate growth
- ◆ Permit movement
- ◆ Provide structural support and rigidity

There are two basic categories of joints in the body: *nonsynovial* and *synovial*. Synovial joints have a space between the bones, which allows free mobility, whereas nonsynovial joints lack such a space and permit limited or no movement between bones. The following sections tell you more about these two categories of joints.

Nonsynovial Joints

Nonsynovial joints are called *synarthroses* or solid joints. In these joints, there is no free space between the bones. Instead, the bones are joined by strong connective tissues, either fibrous connective tissue or cartilage. Movement at these joints is very limited. The type of tissue connecting the bony surfaces describes two categories of synarthroses: *fibrous joints* and *cartilaginous joints*.

Fibrous joints have dense connective tissue between the two bones that could be likened to nylon mesh material. Examples of these include *sutural* joints between the bones of the skull, and *syndesmoses* —the ligaments connecting the two bones in the forearm (the radius and the ulna) and the tibia and fibula in the leg.

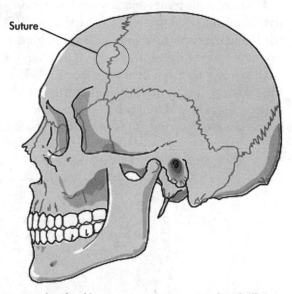

Suture

Example of a fibrous joint, a suture (in the skull).

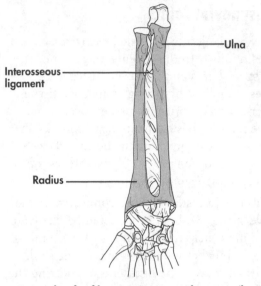

Ulna

Interosseous ligament

Radius

Example of a fibrous joint, a syndesmosis (here between the two bones of the forearm).

Cartilaginous joints occur when the connective tissue between the bones is cartilage (the stuff your ears and nose are made of). There are two types of cartilaginous joints, depending on what type of cartilage they are made of. When the cartilage is hyaline cartilage (a firm but pliable cartilage found on growth plates), the joint is called a *synchondrosis*. This type of joint occurs during longitudinal growth of bone, such as the growth plate cartilage. This cartilage eventually disappears and is replaced by bone. *Symphysis* joints occur where the connective tissue is fibrous cartilage (or *fibrocartilage*). Examples of these joints include the discs between the vertebrae and the joint between the two parts of the pubic bone.

Synovial Joints

Synovial joints, also called *diarthrodial joints*, are the most common joints in the body and permit the greatest degree of movement. The greatest ranges of motions are produced in the ball-and-socket synovial joints of the shoulder and hip, where a person can freely rotate the upper or lower limb around a 360-degree axis (but, thankfully, your shoulder joint rotates more freely than your hip does—unless you're Elvis). The knee and elbow joints are other good examples of synovial joints, but these joints are primarily hinge-type and can move in just two directions: flexed and extended. Both of these joints have some slight rotation. The bones of the wrist, hand, ankle, and foot are also connected to one another with synovial joints.

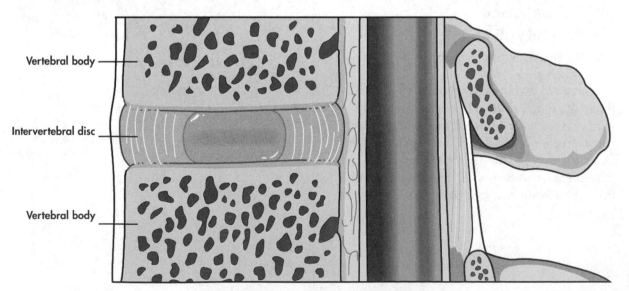

Vertebral body

Intervertebral disc

Vertebral body

Features of a cartilaginous symphysis joint.

The typical synovial joint consists of a joint space or cavity between the two bone surfaces that is enclosed in a sleeve of dense connective tissue fibers, called the *joint capsule*. This capsule is strengthened and reinforced by ligaments. The articular surfaces of the joint, where opposing bones meet and move, have a thick layer of hyaline cartilage, called *articular cartilage*, that is firmly attached to the underlying bone. This cartilage is designed to bear weight. Articular cartilage is unique because, once it is formed, cells within it do not divide, nor does its matrix become mineralized.

Bodily Malfunctions

Because articular cartilage's capacity to repair itself following injury is limited, disease or injury that destroys this cartilage exposes the underlying bone, resulting in pain and stiffness when a person moves the joint. This condition is known as osteoarthritis, or just arthritis.

The inner surface of the joint capsule is lined by the soft tissue of the *synovial membrane*, which is reflected onto and attached to nonarticular surfaces of the joint cavity. This membrane secretes small amounts of *synovial fluid* into the joint space. This fluid is primarily plasma, minus most of its proteins, and is rich in *hyaluronic acid* (which coats each cell and makes the matrix more resilient). Synovial fluid is viscous and very slippery (it sort of looks and feels like egg whites) and functions to moisten and lubricate the articular surfaces to facilitate smooth joint movement. It is also the source of nutrients and metabolic exchange to keep the cartilage healthy.

Body Language

Synovial fluid is like egg white and covers the cartilage in the joint to facilitate smooth movement and keep it healthy.

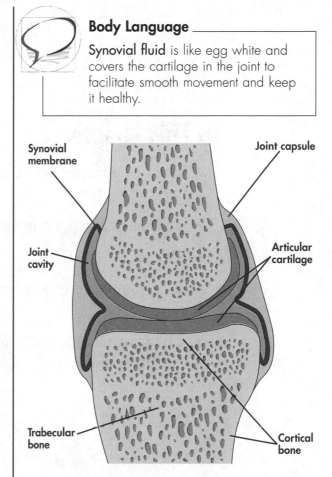

Features of a synovial joint.

The Least You Need to Know

- Bones are living structures containing growing and changing populations of bone cells that are responsible for maintaining bone health.
- The axial skeleton is the central axis of bones in the body that protect the brain, spinal cord, and vital thoracic organs, and to which the appendicular skeleton attaches. The appendicular skeleton contains the bones of the limbs, to which the muscles attach.
- The joints of the body hold the bones together and enable a person to move.

In This Chapter

- The organization and components of the central nervous system

- The parts of the brain and principal features of each

- The spinal cord

- The meningeal coverings of the central nervous system

- The ventricles of the brain and the production of cerebrospinal fluid

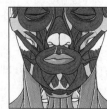
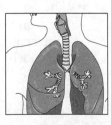
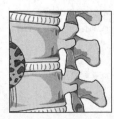
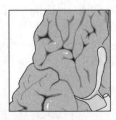

Chapter 5

Grand Central Station: The Central Nervous System

The nervous system is incredibly complex and sophisticated. It simultaneously functions on multiple levels. Through a body-wide network of central and peripheral nerve elements, it enables people to perceive, experience, and interact with their surroundings and to appropriately respond to the external and internal stimuli they receive. These stimuli are relayed to higher centers for information processing, storage, and retrieval; integration; and interpretation, in order to coordinate an appropriate response. In addition, humans' ability to think, reflect, express ourselves through language, and perform high-level mental activities defines who we are (for better or worse) as a species and as individuals.

The nervous system is far too complex to cover completely in a book such as this; however, I hope to give you a framework for understanding its organization and principal activities, which can serve as a step-off point for further reading in this area. In this chapter I talk about the central nervous system (CNS)—that part of the nervous system that includes the brain and the spinal cord. In the next chapter, I discuss the peripheral nervous system, consisting of cranial and spinal nerves, and the role it plays in delivering and receiving information to and from the CNS.

The Brain Trust: Parts of the Brain

The cranial (or head) end of the central nervous system is the brain. The brain is the centerpiece of what makes people human. Our personalities are expressed through this soft tissue structure. Because the brain is delicate and vulnerable, it is protected within a bony box known as the skull. You might be surprised to learn that the brain is not a completely solid structure. Internally, it contains fluid-filled spaces called *ventricles*, which are discussed later in this chapter.

The brain consists of six major parts: the cerebrum, the diencephalon, the cerebellum, and the three parts of the brain stem: the midbrain, the pons, and the medulla oblongata. Each of these parts is discussed in a later section.

Bodily Malfunctions

Hydrocephalus, sometimes referred to as "water on the brain," is a serious disorder that affects 1 in every 500 children. This condition commonly develops from a blockage of the ventricular system or a disruption in the absorption of CSF into the bloodstream. This accumulation of excess fluid causes the ventricles to enlarge, which presses the brain against the inside of the skull. This compression can lead to impaired brain function.

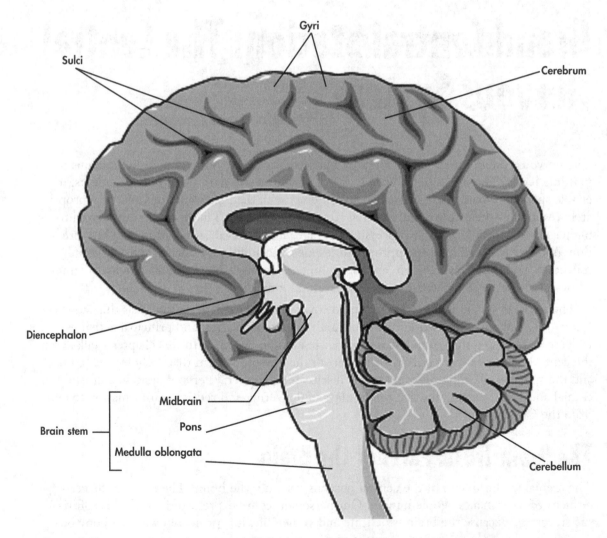

Sagittal view of the brain.

Cerebrum: Where the Real Thinking Happens

The cerebrum represents the largest part of the brain. It's divided into two halves, the right and left *cerebral hemispheres*. The surfaces of these hemispheres consist of a series of ridges or folds, called *gyri* (singular, *gyrus*) and shallow grooves called *sulci* (singular, *sulcus*). Deeper grooves, called *fissures*, separate the two hemispheres.

Foot Notes _____

In the cerebrum, the gray matter is on the outside and the white matter is on the inside. In the brain stem and spinal cord, the gray matter is on the inside and the white matter is on the outside.

Internally, the hemispheres are organized into two kinds of tissue. *Gray matter* consists of cell bodies of neurons and associated neuroglial cells. It's located at the periphery or *cortex* of the cerebrum (like the hard candy part of a Tootsie Roll Pop). *White matter* is located inside the gray matter (like the Tootsie Roll part). White matter consists of large bundles of myelinated nerve fibers. These fibers connect the cerebral cortex to the cerebellar cortex. They also connect the two cortices to lower centers of the brain and spinal cord. Deep within the cerebral hemispheres are collections of nerve cell bodies that form the *basal nuclei*.

Body Language _____

Ganglia are a collection of nerve cell bodies *outside* the central nervous system. *Nuclei* are collections of nerve cell bodies located *inside* the central nervous system. Thus, the traditional term *basal ganglia* is a misnomer. **Basal nuclei** is the correct term for the collection of nerve cell bodies inside the cerebral hemispheres.

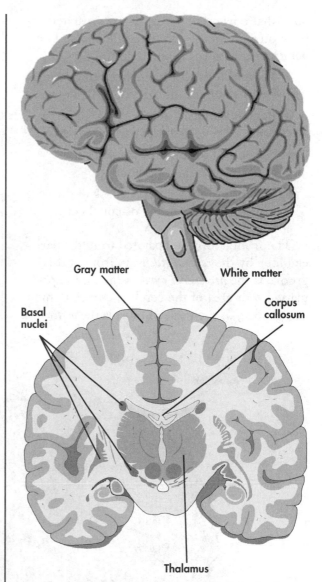

Gray matter   White matter

Basal nuclei   Corpus callosum

Thalamus

The gray matter and white matter of the brain.

The right and left hemispheres are separated by a deep groove called the *longitudinal cerebral fissure*. But the hemispheres are connected at the bottom of this fissure by the *corpus callosum*. The corpus callosum is a thick band of nearly 250 million axons that connects the right and left sides of the cerebrum. It helps coordinate and share information and activities between each side.

Each hemisphere is divided into four lobes. Each lobe corresponds in name with the skull bone that covers it. The lobes are separated from one another by grooves called *sulci*. The names of the lobes are the following:

- ◆ **Frontal lobes.** At the front of the cerebrum and beneath the frontal bone.

- ◆ **Parietal lobes.** At the top of the cerebrum and beneath the parietal bones.

- ◆ **Temporal lobes.** At the sides and beneath the temporal bones.

- ◆ **Occipital lobes.** At the back of the cerebrum and beneath the occipital bones.

The frontal lobe is separated from the parietal lobe by the *central sulcus*. In front of this groove is the *precentral gyrus*, which is the primary motor area of the cerebral cortex. Behind this groove, in the parietal lobe, is the *postcentral*

gyrus, which is the primary sensory area of the cerebral cortex. The *lateral sulcus* separates the temporal lobe from the frontal and parietal lobes. A *parieto-occipital sulcus* marks the boundary between the parietal and occipital lobes.

Within each lobe are primary and secondary *sensory*, *motor*, and *association areas* that support certain functions and activities. Many of these are shown in the following figure.

Body Language

Sensory areas of the brain receive signals from the spinal cord and brain stem regarding touch, taste, smell, sight, and hearing. **Motor areas** initiate messages for movement. **Association areas** are used for more complex mental functions such as recognizing objects.

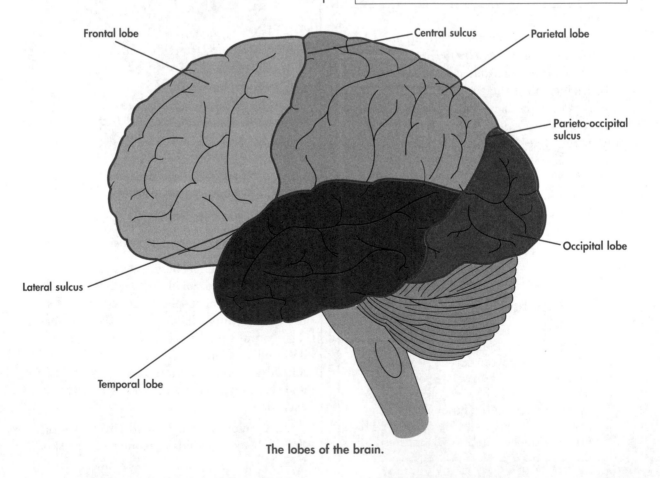

The lobes of the brain.

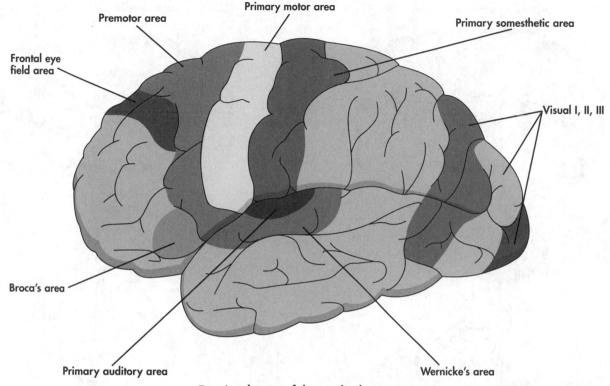

Functional areas of the cerebral cortex.

Body Language

Ascending pathways are nerve highways that send sensory information up to the brain from the body.
Descending pathways send movement instructions from the brain to the rest of the body.

Commands to move muscles in one's *right* arm are initiated from the primary motor area of the *left* cerebral cortex. Similarly, the sense of touch from one side of the body is delivered to the cerebral cortex on the opposite side. This crossover of sensory and motor information occurs at different levels within the central nervous system.

But these activities are, in fact, far more complex than these simple examples. They involve communication and feedback between secondary and association areas. How many muscle fibers need to contract to move my arm? What direction do I want my arm to go? How fast does my arm have to move to catch the ball being thrown to me? Or what was the quality of touch to my skin? Was it painful or not? Did it remind me of the needle stick I got at my last visit to the doctor, the scrape I suffered when I caught the sharp edge of my boss's desk, or a person's finger tapping me on my shoulder? So you see, even the most "simple" activities involve multiple, complex levels of information processing at lightning speeds without us thinking much about them.

Ascending (sensory) pathway

Descending (motor) pathway

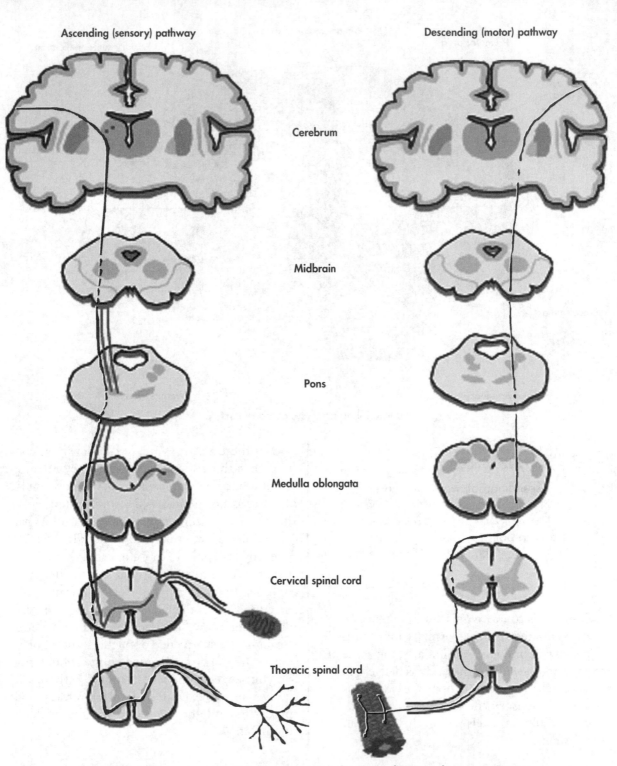

Cerebrum

Midbrain

Pons

Medulla oblongata

Cervical spinal cord

Thoracic spinal cord

Ascending (sensory) and descending (motor) pathways in the central nervous system.

Located deep within the cerebral hemispheres are masses of nerve cell bodies called the *basal nuclei*, or *basal ganglia*. These nuclei receive information from the motor cortex and other areas. They act as reflex centers to control the tone and activity of skeletal muscle. They also work closely with the cerebellum to coordinate movements of the body. These two systems balance each other so that movements are performed smoothly and precisely. Damage to either the basal nuclei or cerebellum will cause some sort of movement disorder.

The Diencephalon

The *diencephalon* is located between the cerebrum and brain stem. The two main parts of it are the *thalamus* and *hypothalamus*. The thalamus is the largest part of the diencephalon. It is divided into right and left halves. The thalamus consists of several nuclei. It's the principal relay station for sensory and motor information between the cerebral cortex and the basal ganglia, cerebellum, brain stem, and spinal cord. It's also involved in hearing and sight system pathways.

Bodily Malfunctions

The role of basal ganglia in the control of muscle activity has been likened to car brakes. They halt unwanted movements. Two movement disorders, Parkinson's disease and Huntington's disease, result from damage or malfunction of these ganglia. Huntington's disease is characterized by uncontrolled, unwanted movements ("no brakes"). One of the manifestations of Parkinson's disease is difficulty in initiating movements ("sticky brakes").

Body Language

The **thalamus** acts as a translator between various parts of the brain. It also regulates states of sleep and wakefulness. The **hypothalamus** regulates body temperature, appetite, and hormonal secretions.

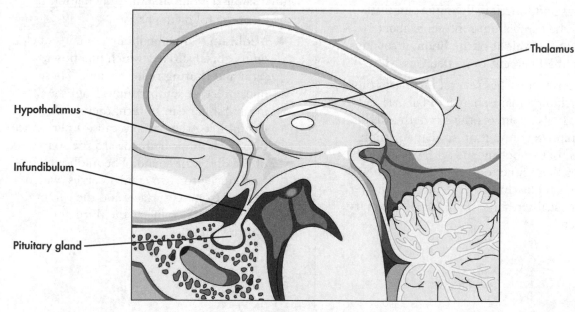

Hypothalamus

Infundibulum

Pituitary gland

Thalamus

The diencephalon.

The hypothalamus is located below and in front of the thalamus and is about the size of an almond. It contains several small nuclei. The hypothalamus performs important functions such as regulation of body temperature and control of appetite. It connects the nervous system to the endocrine system and controls hormonal secretion by the *pituitary gland*. The pituitary gland, or *hypophysis*, is connected by a stalk (the *infundibulum*) to the hypothalamus.

Cerebellum: Chewing Gum and Walking at the Same Time

The *cerebellum* (which means "little brain") is located beneath the occipital lobes of the cerebrum. The cerebellum, like the cerebrum, has a pair of hemispheres, which are connected by the *vermis*. The cerebellum's surface area is increased by a series of narrow ridges, called *folia*, and intervening grooves, called *fissures*.

Internally, each cerebellar hemisphere is organized into an outer cortex of gray matter and a central region of white matter. Deep within the white matter are small masses of *cerebellar nuclei*. The cerebellar nerves are so numerous and functionally important that, whereas the cerebellum represents about 10 percent of the weight of the brain, it contains more than 50 percent of its neurons.

The cerebellum has several important functions, including maintenance of balance and posture. It also controls the strength and duration of muscle contraction so that an action is carried out smoothly and with precision. It performs these functions in collaboration with the basal ganglia. Injury or damage to the cerebellum usually results in abnormal voluntary movement.

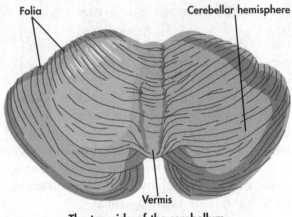

The top side of the cerebellum.

The Brain Stem: Automatic Pilot

The brain stem is located in the center of the brain beneath the cerebral hemispheres. It's a more primitive part of the brain. Nonetheless, it contains vital functional centers that control breathing, blood pressure, and heart rate. In addition, all sensory and motor information passing to and from the cerebral cortex goes through the brain stem.

The brain stem is similar to the spinal cord in having gray matter dispersed centrally and white matter surrounding the gray matter. It consists of the following parts:

◆ **Midbrain.** The midbrain contains nuclei on its dorsal surface, which function as sight and hearing reflex centers. These nuclei receive sensory input from the retina and inner ear. In turn, they send axons to multiple other sites to cause a person to turn his head or body toward the source of the object or sound. The midbrain contains nuclei for two cranial nerves, which emerge from its surface, and the cerebral aqueduct that connects the third and fourth ventricles.

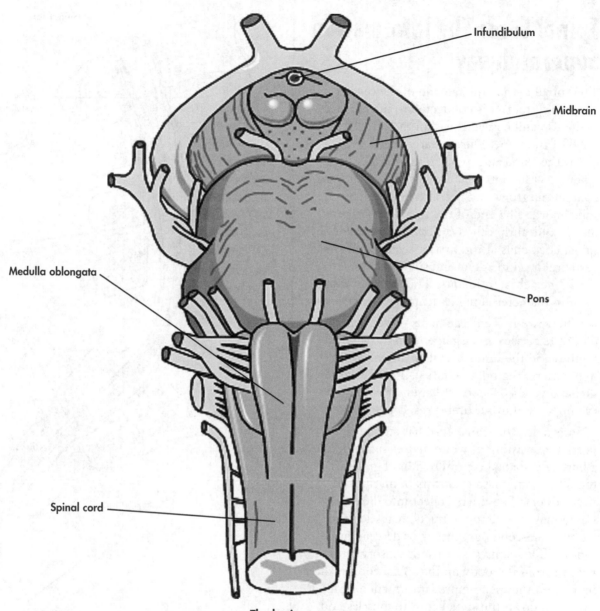

The brain stem.

◆ **Pons.** The pons (meaning "bridge") lies between the midbrain and the medulla. It has a prominent bulge on its front surface and delicate transverse ridges that connect to the cerebellar hemispheres. The pons contains nuclei for four cranial nerves and other nuclei that are involved in regulating breathing.

◆ **Medulla oblongata.** The medulla oblongata connects to the spinal cord at the foramen magnum, the large opening in the base of the skull. It contains nuclei for the last five cranial nerves and other nuclei that help regulate heart rate and breathing.

Spinal Cord: The Information Superhighway

The spinal cord is the tail end of the central nervous system. It is connected to the medulla at the top and descends within the vertebral canal of the spine. During early fetal development, the spine and spinal cord are the same length. With continued development, the spinal column outgrows and becomes longer than the spinal cord. The end of the spinal cord tapers into a conical tip called the *conus medullaris*. The spinal cord ends at the first or second lumbar vertebral level in adults and slightly lower, at the L3 vertebra, in infants. The cord is not uniform in diameter along its length.

There are two enlargements on the spinal cord: the cervical enlargement (in the cervical region) and the lumbosacral enlargement (in the lumbosacral region, of course). These enlargements give rise to nerves that enter and supply the upper and lower limbs, respectively.

Internally, the spinal cord has an outside portion consisting of white matter surrounding a butterfly-shaped central region of gray matter. The white matter consists of myelinated nerve fibers. The fibers collect into different functional and anatomic tracts that belong to pathways (ascending, sensory or descending, motor). The pathways transmit sensory information from the body up the spinal cord to the brain; and they send motor information from the brain down the spinal cord to muscles and other effector tissues.

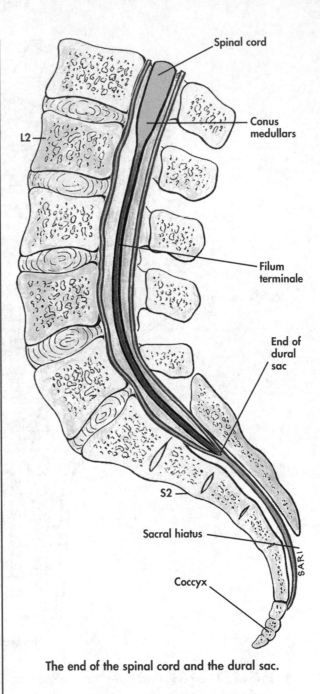

The end of the spinal cord and the dural sac.

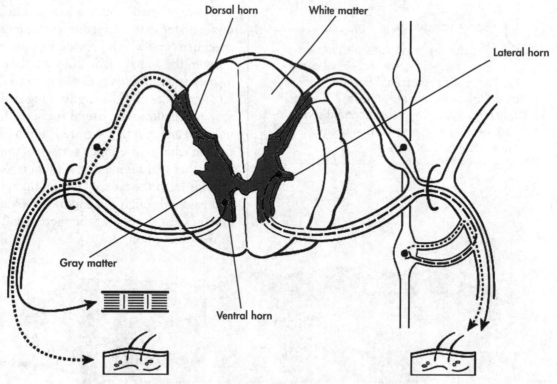

A cross-section of the spinal cord.

Meninges: They've Got You Covered

The brain, spinal cord, and nerve roots are surrounded by three protective membrane layers, collectively called the meninges. These layers include the following:

◆ **Dura mater (the "tough mother").** This is the outermost covering. It is composed of tough, fibrous connective tissue. It attaches to the inner surface of the skull and continues below within the vertebral canal of the spine as an elongated sac that encloses the spinal cord, called the *dural sac*. This sac ends at the S2 vertebral level. Spinal nerves must pierce this sac to exit the vertebral canal.

◆ **Arachnoid mater (the "spider-web" layer).** This is a delicate, fibrous membrane that is loosely attached to the inner surface of the dura. It is held in place by the pressure of the *cerebrospinal fluid (CSF)* beneath this layer.

◆ **Pia mater (the "tender or faithful" one).** This is the innermost membrane covering. It is very thin and firmly attached to the surfaces of the brain, spinal cord, and nerve roots. The arachnoid mater and pia mater are separated by a fluid-filled space called the *subarachnoid space*. This space contains cerebrospinal fluid, which cushions and protects the brain and spinal cord.

Body Language

The **dural sac** is the tissue surrounding the spinal cord and spinal nerve roots. **Cerebrospinal fluid** is the clear liquid that cushions and protects the brain and spinal cord. The **subarachnoid space** is between two layers of the meninges and contains cerebrospinal fluid.

There are two modifications or thickenings of pia mater that are readily visible in gross specimens and which provide support and stability to the spinal cord. They are the *denticulate ligaments* and the *filum terminale*. The denticulate ligaments are white, glistening structures that attach along the lateral surfaces of the spinal cord and end in a serrated "tooth-like" edge that attaches to the inner surface of the dura. The other pial remnant, the filum terminale, extends from the tapered end of the spinal cord and passes through the end of the *dural sac*. It emerges out the end of the sacrum and attaches to the coccyx bone.

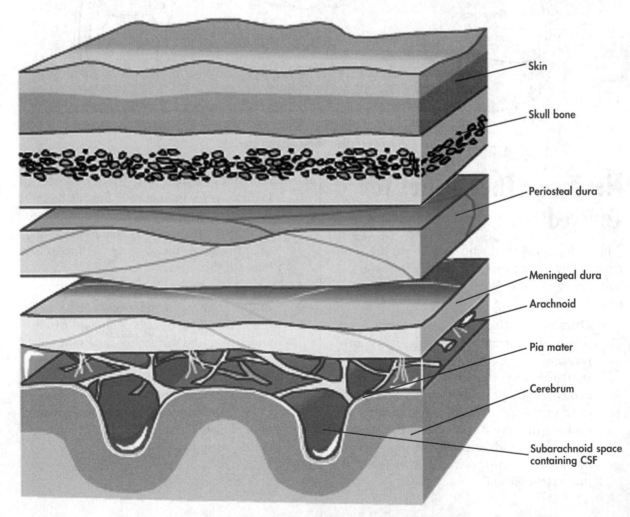

Skin

Skull bone

Periosteal dura

Meningeal dura

Arachnoid

Pia mater

Cerebrum

Subarachnoid space containing CSF

Cross-section of skull, meninges, and surface of the brain.

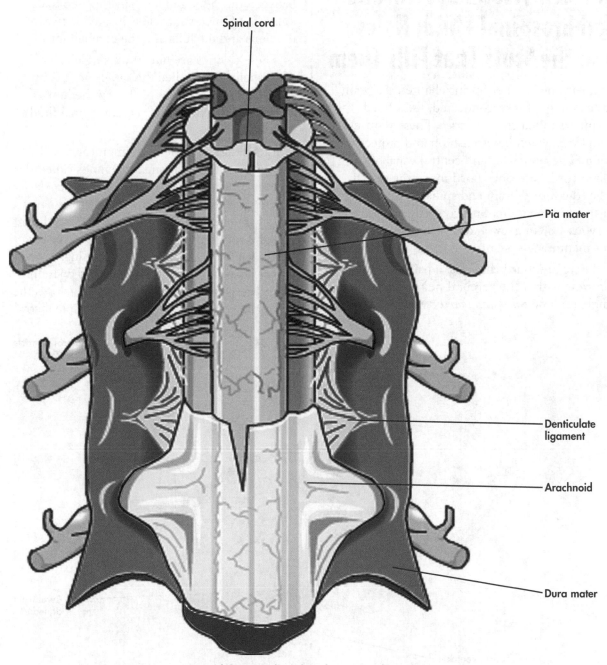

View of the spinal cord and meningeal coverings.

The Ventricular System and Cerebrospinal Fluid: Holes and the Stuff That Fills Them

As mentioned earlier in this chapter, the brain has a system of interconnected, fluid-filled chambers called the *ventricles*. These ventricles hold CSF, which helps cushion and protect the brain. The ventricles and central canal of the spinal cord were once portions of the neural tube (the future brain and spinal cord) during early embryonic life. Subsequently, the central nervous system grew and developed external to this membrane system.

There are four ventricles: a pair of *lateral ventricles* within the cerebral hemispheres, and single midline *third and fourth ventricles*. The third ventricle is located in the diencephalon between the right and left halves of the thalamus. The fourth ventricle is located between the pons and medulla and the cerebellum.

These cavities are lined by a single layer of epithelial cells called the *ependyma*. Within each ventricle, portions of this epithelium are involved in producing cerebrospinal fluid (CSF).

Each lateral ventricle communicates with the third ventricle via small openings called the *interventricular foramina*. The third ventricle is connected to the fourth ventricle by a narrow channel called the *cerebral aqueduct*, which passes through the midbrain. The roof of the fourth ventricle contains three openings (two lateral holes and a midline hole) that open into the subarachnoid space. This space encloses the brain and continues down along the spinal cord.

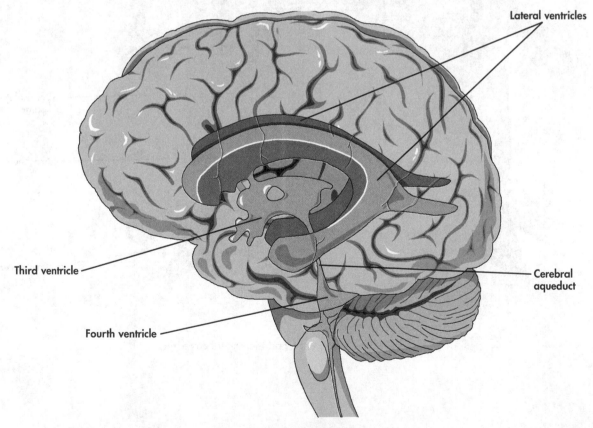

Lateral ventricles

Third ventricle

Cerebral aqueduct

Fourth ventricle

The ventricles of the brain.

The ventricular system is constantly producing CSF, which is also constantly being absorbed by the venous system surrounding the brain (more on this in Chapter 21). Blockage along the ventricular system or disruption of CSF reabsorption can lead to the condition known as hydrocephalus (or water on the brain).

The Least You Need to Know

◆ The central nervous system consists of the brain and spinal cord.

◆ All sensory information about the external and internal environments of the body is communicated to the brain along ascending pathways in the spinal cord.

◆ Motor pathways descend from the brain and pass down the spinal cord to cause contraction of muscles.

◆ The meninges consist of three layers of non-nervous tissue and enclose and protect the brain and spinal cord.

◆ Ventricles within the brain are cavities that contain and produce CSF, which is circulated around the brain and spinal cord to protect it.

In This Chapter

- ◆ The difference between the CNS and the PNS
- ◆ The difference between a spinal nerve and a cranial nerve
- ◆ How the somatic nervous system works
- ◆ How the autonomic nervous system works

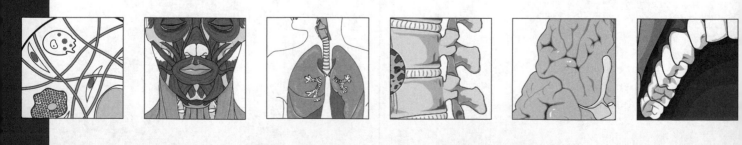

Chapter 6

Left of Center: The Peripheral Nervous System

In the preceding chapter, I introduced the central nervous system, consisting of the brain and the spinal cord. The CNS receives, decodes, processes, and responds to information it receives from the body, by means of the *peripheral nervous system (PNS)*. Information passes up and down the CNS via bundles of nerve fibers called tracts or pathways. Nerve fibers in the PNS are bundled together and called nerves.

The PNS has a lot of nerves. But don't be nervous. I'll try to package this information in a manner that makes it easy to understand.

Full of Nerves: The Many Aspects of the PNS

The PNS can be described in a couple different ways. Anatomically, it consists of 31 pairs of spinal nerves, so called because they arise from the spinal cord, and 12 pairs of cranial nerves that originate from the brain.

Functionally, the PNS is divided into two nervous systems, the *somatic nervous system (SNS)* and the *autonomic nervous system (ANS)*. Both systems share some of the same "nerve highways." The SNS is comprised of somatic sensory and somatic motor neurons. These neurons relay feelings from the skin, muscles, tendons, and joints relating to pain, touch, pressure, temperature, tone, stretch, or position sense. They also supply skeletal muscle. Each sensory or motor nerve fiber involved in these functions represents just one nerve cell from the CNS to its "target" tissue. This means that a motor neuron supplying a muscle in the big toe may be a few feet in length from its origin in the spinal cord! The SNS is a voluntary system because a person has conscious control over the movement of the skeletal muscles.

Body Language

The **somatic nervous system** relays feelings from the body's external environment to the central nervous system. "Soma" is Greek for "body. The **autonomic nervous system** supplies nerves to the internal organs. "Autonomic" means, roughly, "not under voluntary control."

The autonomic nervous system, on the other hand, largely supplies and regulates the body's internal organs (also called *viscera*). It is the system that controls the cardiovascular, respiratory, digestive, urinary, reproductive, and endocrine systems. The neurons responsible for these activities are known as *visceral sensory* and *visceral motor neurons*. Because a person cannot control these activities with his or her will, the ANS is called an involuntary system. The specific "target" tissues of these neurons are smooth muscle, cardiac or heart muscle, and glands. Unlike in the SNS, it takes a two-nerve chain of visceral motor neurons to activate these target tissues.

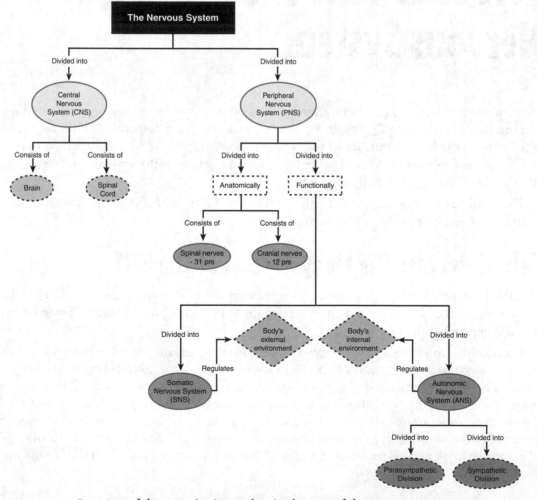

Overview of the organization and main elements of the nervous system.

The ANS is further separated into two divisions:

◆ **The sympathetic division.** The "fight or flight" system. This is probably the system that is working when you give your first public speech.

◆ **The parasympathetic division.** The vegetative "rest and digest" system. You feel this system in action after your nice big Thanksgiving meal.

Although these systems have opposite purposes, they work together cooperatively to keep the organ systems stable. You can think of the sympathetic nervous system as the gas pedal and the parasympathetic nervous system as the brake.

Spinal Nerves: The Messengers

A spinal nerve is so named because it is attached to the spinal cord. Two roots, a dorsal root and a ventral root, enter and leave each side of the spinal cord at regular intervals. A dorsal root attaches to a dorsal horn and a ventral root attaches to a ventral horn. Ventral roots contain mostly axons of somatic motor neurons. However, some ventral roots also contain axons of preganglionic autonomic neurons. These motor fibers are *efferent fibers* and send information away from the spinal cord.

Body Language

Motor axons that send information away from the brain and spinal cord are called **efferent fibers.** Sensory neurons that send impulses toward the spinal cord and brain do so via nerve processes called afferent fibers.

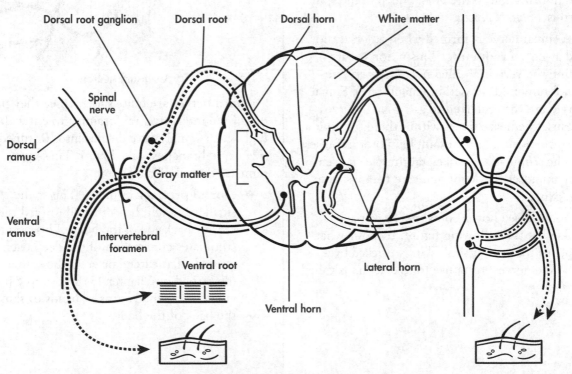

Cross-section of the spinal cord and spinal nerve formation.

The Somatic Component of Spinal Nerves

Dorsal roots contain nerve fibers of somatic and visceral sensory neurons. These sensory fibers send impulses toward the spinal cord. Along each dorsal root is a swelling called the *dorsal root ganglion* or *spinal ganglion*. This swelling is produced by collections of nerve cell bodies of sensory neurons.

These nerve cells look different than motor neurons. Instead of having a cell body located at one end of its axon, its body is located off to the side of its axon and connects to it like an upside-down "T." It has a single process, which acts like an axon and has two parts. The part that extends out into the body and whose ending is specialized to respond to sensory stimuli (for example, feelings of pain, touch, pressure, temperature, or dilation, constriction, or reduced blood supply to internal organs) is called a *peripheral process*. The part that conveys this information to the dorsal horn of the spinal cord is a *central process*.

A spinal nerve is formed when a dorsal root and a ventral root unite. This union occurs within the vertebral canal near an intervertebral foramen (discussed in Chapter 15). Spinal nerves are thus referred to as "mixed" nerves because the nerve fibers within them transmit a mixture of signals; for example, motor impulses for the stimulation of skeletal muscle and sensory impulses that communicate feelings from the skin.

You can see how a spinal nerve works by looking at the following figure. It shows what happens when a person's skin is pricked by a pin. The nerves' response to this action is called a *reflex arc*.

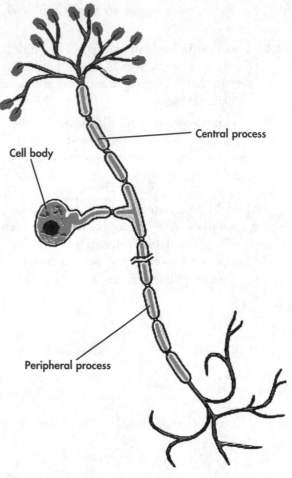

A sensory neuron.

Spinal nerves are quite short. Once they pass out of the intervertebral foramina to enter the body, each spinal nerve divides into two unequal size nerve branches, called rami. These two rami are described as follows:

◆ **Dorsal primary ramus.** This is the smaller branch. It is directed toward the rear or backside of the body. This branch stimulates contraction of the deep back muscles that extend or erect the spine (described in Chapter 15) and supply individual horizontal strips or bands of skin at the back of the body.

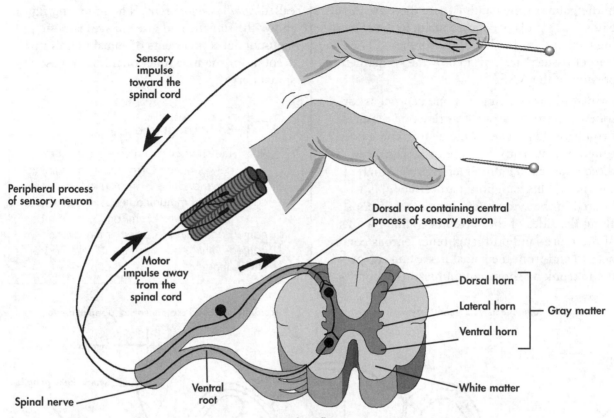

Sensory
impulse
toward the
spinal cord

Peripheral process
of sensory neuron

Motor
impulse away
from the
spinal cord

Spinal nerve

Ventral
root

Dorsal root containing central
process of sensory neuron

Dorsal horn

Lateral horn — Gray matter

Ventral horn

White matter

A simple reflex arc.

◆ **Ventral primary ramus.** This is the larger branch. Nerve fibers within it provide motor supply to muscles of the body wall and upper and lower limbs, and to more extensive areas of skin.

Foot Notes

The skin surface of the body can be mapped into a series of bands or areas, referred to as dermatomes. A dermatome is a specific area of skin supplied by a single spinal nerve or spinal cord level. For example, the skin around the belly button is supplied by the tenth thoracic spinal nerve. Loss or disturbance of feeling to an area of skin may indicate a problem with the nerve supplying that portion of skin or some problem involving the spinal cord or higher center within the CNS.

The Autonomic Component of Spinal Nerves

So far I've limited my description of spinal nerves to the principal nerve fiber components within them, the somatic sensory and motor neurons, and the tissues they supply (for example, skeletal muscle and skin). But they also contain nerve fibers associated with the ANS, specifically, the sympathetic division.

Have you ever had "goose bumps" on your skin while shivering from the cold or fright? Or sweated a lot from some stressful situation? In Chapter 3 you learned that connected to the hair follicles in the skin were tiny, smooth muscles called arrector pili, which when contacted, make the hair stand on end. Also the skin contains sweat glands—lots of them. They are widespread throughout the body, but particularly numerous in the palms of the hands (where there are about 650 sweat glands per square inch), in the soles of the feet, and in the skin of the forehead. These "target tissues" are activated by the sympathetic portion of the ANS.

Instead of requiring just one neuron, as for somatic motor neurons, this function requires two nerves (a preganglionic neuron and a post-ganglionic neuron). Synapse occurs between these neurons in an intervening sympathetic ganglion. This ganglion (paravertebral) is located in the sympathetic trunk that extends along the sides of the vertebral column. Axons of these pre- and postganglionic neurons connect to and from the spinal nerve and sympathetic trunk by short communicating branches

called *rami communicantes*. The following figure shows the somatic and visceral sympathetic (autonomic) components of spinal nerves that account for the four functional fiber types in spinal nerves.

Body Language

The short communicating branches between the ventral ramus of a spinal nerve and the sympathetic trunk are called **rami communicantes**. There are two types of these: white rami and gray rami, named according to whether the axons carried within these rami are myelinated or unmyelinated.

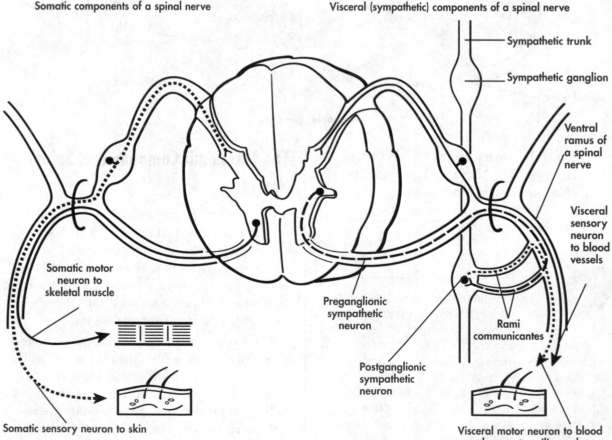

Somatic and visceral (sympathetic) components of spinal nerves.

Think of an electrical wire that contains within it four smaller colored wires. The big wire is the "nerve" as you would see it in a dissection or operation. The four smaller wires represent the different functional types of fibers in the nerve:

◆ **Somatic motor.** A single neuron that stimulates skeletal muscle to contract

◆ **Somatic sensory.** A single neuron that carries general sensory information from the skin

◆ **Visceral motor.** A two-neuron chain that stimulates smooth muscle contraction or sweat gland secretion

◆ **Visceral sensory.** A single neuron that carries visceral sensory information from blood vessels in the skin

A nerve (the electrical wire) that is visible to the human eye represents hundreds to thousands of axons (the small colored wires) contained within a sleeve of connective tissue that transmits these various somatic and visceral sensory and motor impulses between structures of the body and the CNS.

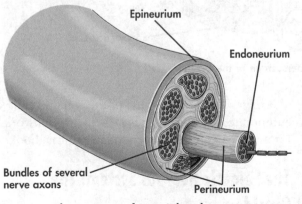

Epineurium

Endoneurium

Bundles of several nerve axons

Perineurium

The anatomy of a peripheral nerve.

Bodily Malfunctions

Severe injury to a peripheral nerve (for example, severing it in a cut) may cause a number of symptoms following injury that relate to the functional fiber types in the nerve. Such an injury can cause paralysis of skeletal muscle, numbness of skin, dry skin (because of loss of nerve supply to sweat glands), and skin that's a bit warmer than normal (because of loss of smooth muscle control in blood vessel walls).

Cranial Nerves: The Special Sensory Connection

Cranial nerves (CN) are the other group of nerves in the peripheral nervous system. There are 12 pairs of cranial nerves that come off the bottom of the brain at irregular intervals.

In addition to having names, cranial nerves are numbered using roman numerals. Unlike spinal nerves, cranial nerves vary in the number of functional fiber types they contain. Some, like the *olfactory nerve* (CN I) or *trochlear nerve* (CN IV), contain only one type, sensory or motor. Others, like the *facial nerve* (CN VII) or *glossopharyngeal nerve* (CN IX) contain several types: sensory, motor, and parasympathetic. Further details regarding these nerves and the structures they supply are given in Chapter 21.

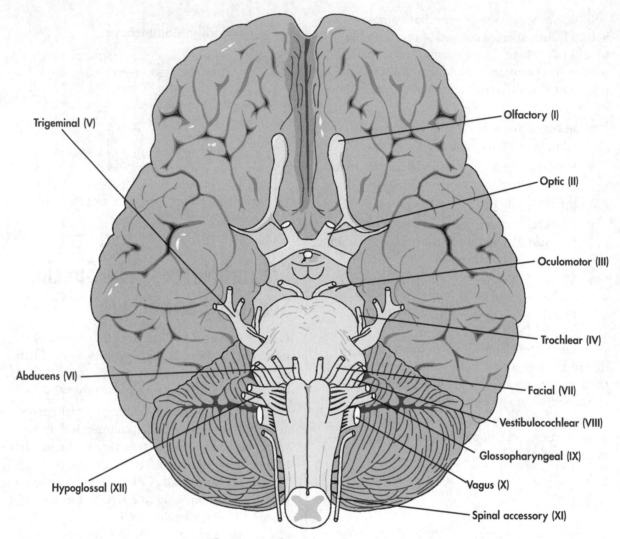

Trigeminal (V)

Olfactory (I)

Optic (II)

Oculomotor (III)

Trochlear (IV)

Abducens (VI)

Facial (VII)

Vestibulocochlear (VIII)

Glossopharyngeal (IX)

Hypoglossal (XII)

Vagus (X)

Spinal accessory (XI)

Cranial nerves on the undersurface of the brain.

It's About Regulation

The nervous system has a tremendous collection of checks and balances to its overall operation so that the body and its parts work smoothly, precisely, and cooperatively. It seems the right hand always knows what the left hand is doing (and vice versa) and the right part of the nervous system is being used for any given situation a person finds himself in. This is good for us!

The following discussion of the somatic and autonomic nervous systems lays the groundwork for a more detailed discussion later in this book.

The Somatic Nervous System: The Outsiders

The SNS is the functional part of the peripheral nervous system that regulates the body's external environment and provides motor impulses to skeletal muscle. This happens through 31 pairs of spinal nerves:

◆ 8 pairs of cervical nerves

◆ 12 pairs of thoracic nerves

◆ 5 pairs of lumbar nerves

◆ 5 pairs of sacral nerves

◆ 1 pair of coccygeal nerves

The functions of spinal nerves are carried out by dorsal and ventral branches (rami). The influence of the dorsal rami in the body is quite limited. They supply motor nerve impulses to columns of deep back muscles that extend on either side of the spine from the back of the neck to the lower lumbar and sacral regions. They also relay sensory information from the skin along the middle of the back.

The rest of the body—the neck, trunk, and limbs—is supplied by ventral rami. This system of nerve supply to the trunk (thoracic and abdominal walls) is quite straightforward. It consists of a segmental series of individual ventral rami that supply muscle and skin. (This is covered further in Chapters 17 and 18.) The nerve supply to the neck and limbs is made a bit more difficult to understand because these nerves originate from *nerve plexuses* that are first formed by the intermingling of fibers from several ventral rami. Here's a brief description of these plexuses (you'll see more detail in later chapters):

◆ **Cervical plexus.** Formed by ventral rami of C1–C4 spinal nerves. The cervical plexus provides the nerve supply to most of the muscles and skin of the neck.

◆ **Brachial plexus.** Formed by ventral rami C5–C8 and T1 spinal nerves. The brachial plexus provides the nerve supply to the muscles and skin of the upper limb.

◆ **Lumbosacral plexus.** Formed by ventral rami of the L2–S3 spinal nerves. This *plexus* provides the nerve supply to the muscles and skin of the lower limb.

Body Language

A **plexus** in the nervous system is a network of intersecting nerves. These are the places in which branches of related nerves come back together into one large nerve group.

The Autonomic Nervous System: Fight or Flight vs. Rest and Digest

The ANS is the functional part of the peripheral nervous system that controls the activity of smooth muscle, cardiac muscle, and glands. This is accomplished by the largely antagonistic but cooperative activity of its two divisions: the sympathetic and parasympathetic divisions. The sympathetic division has a more widespread action in the body, controlling structures in the skin (for example, blood vessels, arrector pili muscles, and sweat glands) as well as the internal organs. The parasympathetic division is mainly concerned with co-regulation of internal organs. Its fibers are not distributed in spinal nerves.

The sympathetic system is the system that in stressful situations prepares the body to take off running or stay and fight. Some of the physiological responses to this body-wide activation include:

◆ Increased heart rate and strength of heartbeat

◆ Dilation of the pupils

◆ Dilation of the bronchi

◆ A rush of adrenaline that makes you feel all wobbly

The parasympathetic system works in the opposite way to return body function back to normal once the danger has passed. Among its effects are a decrease in heart rate and strength of heartbeat, constriction of pupils, and a resumption of digestive activities.

As stated previously, the visceral motor component of the ANS involves a two-neuron chain from the CNS to the target tissue. The cell bodies of the preganglionic neurons are located within cell columns or nuclei within the CNS. The cell bodies of the postganglionic neurons are located in autonomic ganglia within the PNS.

Each division of the autonomic nervous system gets its signals from the CNS in a different way. The sympathetic division has a *thoraco-lumbar outflow*. Its preganglionic cell bodies are found associated with the T1–L2 levels of the spinal cord. The parasympathetic division has a *craniosacral outflow*. Cell bodies of preganglionic parasympathetic neurons are located within the brain stem. Their axons enter only cranial nerves III, VII, IX, and X. Also cell bodies are located in the lateral horns of sacral spinal cord levels S2, 3, and 4.

Body Language

Thoracolumbar outflow is a term that describes the sympathetic division, by reference to the location of the cell bodies of its preganglionic neurons in lateral horns of the T1–L2 segments of the spinal cord. Craniosacral outflow describes the parasympathetic division by reference to the location of the cell bodies of its preganglionic neurons in the brain stem and the S2, S3, and S4 segments of the spinal cord.

The many actions of these two parts of the ANS on body organs are shown in the following figures.

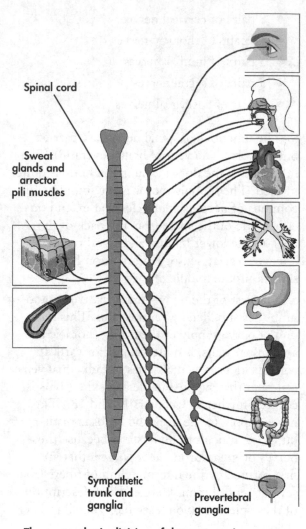

The sympathetic division of the autonomic nervous system.

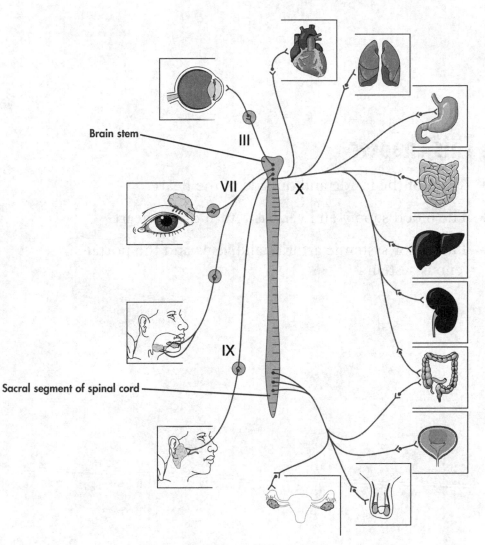

Brain stem

Sacral segment of spinal cord

III

VII

X

IX

The parasympathetic division of the autonomic nervous system.

The Least You Need to Know

◆ The peripheral nervous system consists of 31 pairs of spinal nerves and 12 pairs of cranial nerves.

◆ Spinal nerves contain mostly somatic motor and sensory fibers to skin and skeletal muscles, and some sympathetic fibers that supply sweat glands, blood vessels, and arrector pili muscles.

◆ Cranial nerves contain variable numbers of nerve fiber types.

◆ The somatic nervous system uses single motor neurons from the central nervous system to their target tissue, whereas the autonomic nervous system uses a two-neuron chain.

◆ The sympathetic division of the autonomic nervous system is the system that prepares the body to respond to stress, whereas, the parasympathetic division helps reverse these effects.

In This Chapter

◆ Basics of the inside and outside of the heart

◆ The blood supply and venous return of the heart

◆ The major systemic arteries and veins and the portal venous system

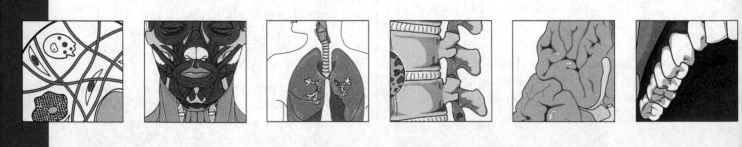

Chapter 7

The Heart of the Matter:
The Cardiovascular System

The "heart" of the cardiovascular system is, of course, the heart, which functions as a pump. The rest of the cardiovascular system consists of an enormous network (tens of thousands of miles, in fact) of blood vessels that transports blood to the organs, tissues, and cells of the body and returns it back to the heart.

Just as the heart lies center stage in love, so too does it in life, literally. The body cannot survive more than a few minutes without a heartbeat. This chapter gives you the basic information about the anatomy of the heart, inside and out, and the major vessel highways.

Imagine doing the same thing once every second, day in and day out, rain or shine, for an entire lifetime. That's what the heart does. About 86,400 times each day, it beats and delivers vital oxygen- and nutrient-rich blood throughout the body. Amazingly, it has been performing this vital task in each human ever since he or she was a 22-day-old embryo! The heart was truly born to beat. You need to know about several important external and internal features of the heart in order to understand how it works. I will describe the external features first.

External Features of the Heart: It's a Cone Thing

The heart is often described as being about the size of a clenched fist. In reality, it is slightly larger than a typical fist. When viewed from the front, the heart is triangular, although a cone shape or toy top is a better three-dimensional description. The heart consists of a *base* and an *apex*. The base is its broader top end, from which the great vessels (discussed later) enter and leave the heart. From the base, the heart narrows to a blunt, rounded point called the apex, which projects downward and toward the left. Within the chest, the bulk of the heart is positioned behind the sternum, and the apex extends a short distance to the left of the sternum (not coincidentally, right where you put your hand when you say the Pledge of Allegiance).

The Groove Is in the Heart

The heart has four chambers: two atria and two ventricles. Two atria are at the base of the heart. Beneath them are the two ventricles that extend toward the apex. On the surface of the heart separating these chambers are visible grooves, called *sulci* (or sulcus, when it's singular). The names of these grooves are pretty easy to remember once you know the atria from the ventricles.

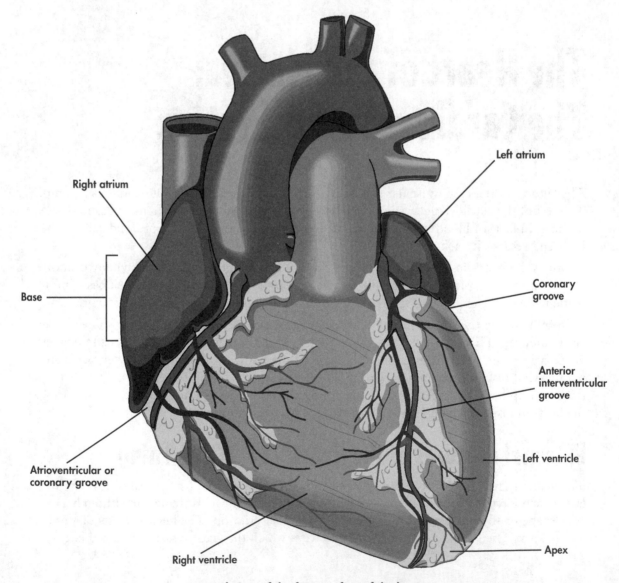

External view of the front surface of the heart.

In addition, some of the heart vessels have names that are similar to the names of the grooves. The groove that separates the atria from the ventricles and completely encircles the heart is the *atrioventricular groove*, also known as the coronary groove because it "crowns" the base of the heart. On the front (anterior) and back (posterior) surfaces between the right and left ventricles are additional shallow grooves called the *anterior and posterior interventricular grooves*, respectively. These grooves are on top of the muscle wall (the interventricular septum) that separates these chambers. Within these grooves are the coronary arteries and their branches, as well as the cardiac veins.

The Sender: Coronary Arteries

The heart has its own mini-circulatory system. It has a pair of *coronary arteries*, a right one and a left one, that supply it with arterial blood. It also has a system of veins, called *cardiac veins*, that return venous blood to the right atrium.

Each coronary artery arises from the beginning of the aorta. (The aorta, which is discussed later in this chapter, sends blood from the left ventricle to the rest of the body.) The right coronary artery courses within the atrioventricular groove between the right atrium and the right ventricle. It gives off branches to portions of these two chambers. The right coronary

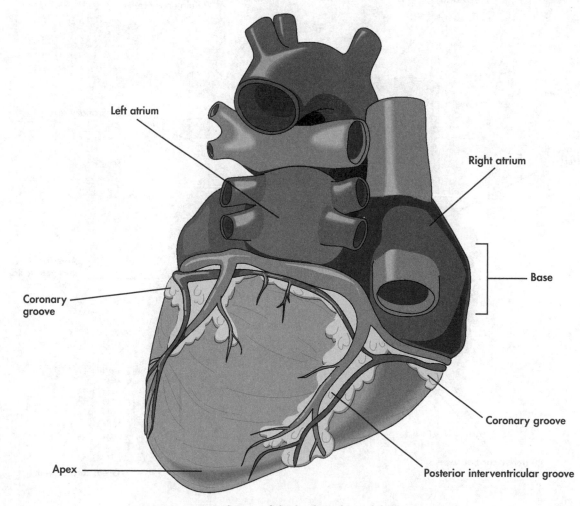

Left atrium

Right atrium

Base

Coronary groove

Coronary groove

Posterior interventricular groove

Apex

External view of the back surface of the heart.

artery continues within this groove onto the backside of the heart. It typically ends by turning downward and coursing into the posterior interventricular groove. When it does this, its name changes to the *posterior interventricular artery*. On this surface of the heart, it provides branches to the posterior portions of the right atrium, the right and left ventricles, and the posterior third of the interventricular septum.

After leaving the aorta, the left coronary artery courses to the left within the atrioventricular groove (which separates the left atrium from the left ventricle) for a short distance before dividing into two branches: the *anterior interventricular artery* and the *circumflex artery*. The anterior interventricular artery, also known clinically as the left anterior descending artery, courses downward within the anterior interventricular groove. It gives rise to branches that supply the major part (the anterior two thirds) of the interventricular septum and adjacent walls of the right and left ventricles. The left circumflex artery continues around to the backside of the heart within the atrioventricular groove. It typically gives off branches to the posterior wall of the left ventricle.

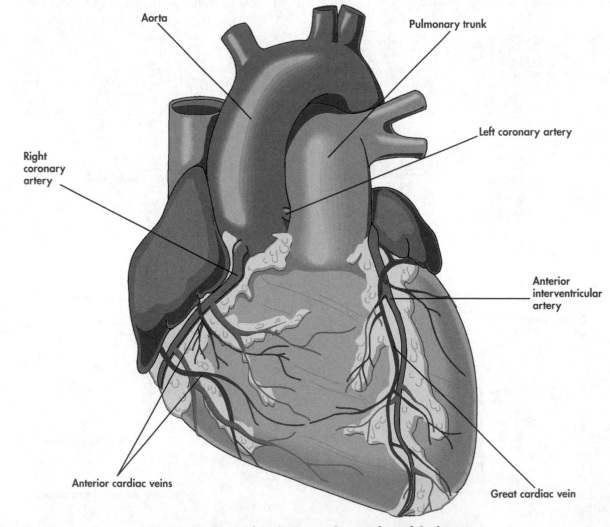

Coronary arteries and cardiac veins, front surface of the heart.

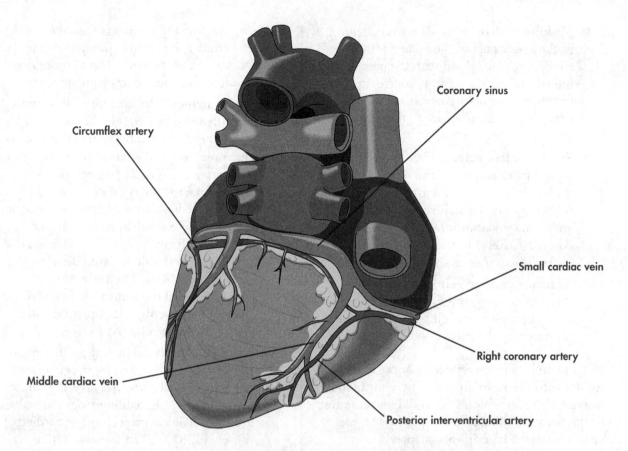

Circumflex artery

Coronary sinus

Small cardiac vein

Right coronary artery

Middle cardiac vein

Posterior interventricular artery

Coronary arteries and cardiac veins, back surface of the heart.

There are a variety of different branching patterns of the coronary arteries, depending on the person, but this is the most common pattern.

Bodily Malfunctions

Atherosclerosis is when fats, cholesterol, and other substances, called plaque, build up in the walls of arteries, causing them to narrow and reduce blood flow to the tissues. Rupture of the plaque or formation of a blood clot may cause a heart attack (myocardial infarction). When blockage of an artery is severe, it may be opened by a metal stent inserted within the artery (by a procedure known as angioplasty) or "bypassed" by attaching a healthy vessel around the site of the blockage.

Return to Sender: Cardiac Veins

The main path for the return of venous blood from the heart is via the cardiac veins into the coronary sinus and the right atrium. There are several cardiac veins:

◆ **Great cardiac vein.** This vein ascends in the anterior interventricular groove, alongside the anterior interventricular artery, to the coronary groove. It follows in this groove to the back side of the heart, where it enters the coronary sinus.

◆ **Coronary sinus.** This is a wide venous channel in the coronary groove on the back of the heart. It receives the great, middle, and small cardiac veins. The coronary sinus ends as an opening into the right atrium.

◆ **Middle cardiac vein.** This vein begins in the lower part of the posterior interventricular groove (along with the posterior interventricular artery). It courses upward in this groove before entering the coronary sinus near its entrance into the right atrium.

◆ **Small cardiac vein.** This vein begins on the front surface of the heart along the lower margin of the right ventricle. It then enters the atrioventricular groove and courses within it (along with the right coronary artery) to the back side of the heart. There it enters the coronary sinus.

◆ **Anterior cardiac veins.** Usually two or three in number, these veins return blood from the right ventricle. They enter the right atrium directly.

Ultimately, venous return from the heart itself is into the right atrium. There the blood mixes with venous blood delivered from the rest of the body via the *superior* and *inferior venae cavae* (discussed later in this chapter).

Internal Features of the Heart: I Have a Partition of Four!

Beneath the external grooves are internal partitions, or walls, that divide the interior of the heart into four chambers. Each of these chambers contains unique structures and features. This section discusses these chambers in detail.

Chambers of Commerce

The heart is hollow and consists of four chambers: two upper chambers called *atria* (a right one and a left one) and two lower chambers called *ventricles* (right and left). The heart functions as a two-sided pump. The right atrium and ventricle work together to pump

oxygen-poor blood to the lungs for oxygenation. The left atrium and ventricle pump oxygen-rich blood to the rest of the body. Here's more detail on the function of these four chambers:

◆ **Right atrium.** Its interior wall is partly smooth and partly rough. The smooth portion receives the openings of the superior vena cava, the inferior vena cava, and the coronary sinus. The rough portion of this chamber is the right *auricle* (an "ear-like" projection). Its wall contains *pectinate muscles*. These are delicate, comb-like ridges of cardiac muscle. The *interatrial septum* is a thin wall of tissue between the right and left atria. The right atrium communicates with the right ventricle through the right atrioventricular opening and its right atrioventricular (AV) valve.

◆ **Right ventricle.** The wall of this chamber is thick and marked by irregular muscle ridges called *trabeculae carneae* ("thick, fleshy beams"). Extending from the walls are short, conical projections of cardiac muscle called *papillary muscles*. There are three of these papillary muscles (anterior, posterior, and septal) in the right ventricle. They correspond in name and number to the cusps of the right AV (or tricuspid) valve. At the tips of the papillary muscles are thin, tough tendons called *chordae tendineae* that connect a papillary muscle to the free edges of valve cusps. The papillary muscles keep the cusps from being pushed into the atria when the ventricles contract. They also maintain a constant tension on the cusps. This chamber opens to the beginning portion of the *pulmonary artery* (also known as the pulmonary trunk) at the pulmonary opening. This is where the pulmonary semilunar valves are located. The right ventricle is a low-pressure pump because it has to push blood only a short distance to the lungs.

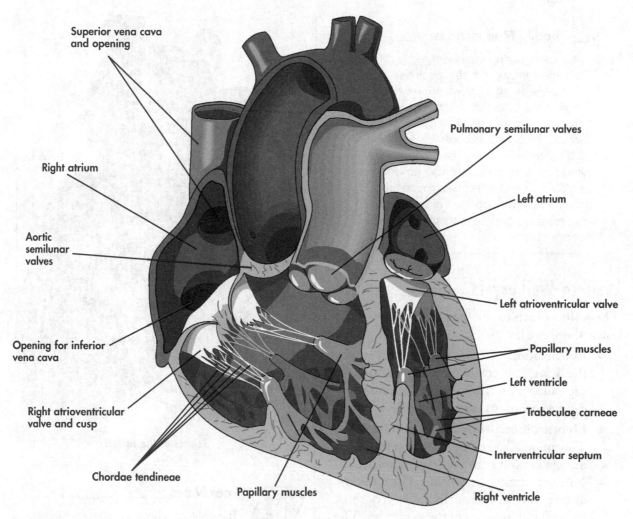

Superior vena cava and opening

Right atrium

Aortic semilunar valves

Opening for inferior vena cava

Right atrioventricular valve and cusp

Chordae tendineae

Papillary muscles

Pulmonary semilunar valves

Left atrium

Left atrioventricular valve

Papillary muscles

Left ventricle

Trabeculae carneae

Interventricular septum

Right ventricle

Chambers of the heart.

◆ **Left atrium.** The wall of this chamber is similar to that in the right atrium. There is a larger smooth portion and a smaller rough portion containing the comb-like pectinate muscles in the left auricle. This chamber receives two pairs (right and left) of pulmonary veins that return oxygenated blood from the lungs. It opens to the left ventricle by the left atrioventricular opening and its left valve.

◆ **Left ventricle.** The wall of this chamber is about three times thicker than the

wall of the right ventricle. This chamber is a high-pressure pump that pushes oxygenated blood to the rest of the body. Its internal surface is roughened by trabeculae carneae. It contains two papillary muscles (anterior and posterior) whose chordae tendineae attach to valves (more on these later). Leading out of the left ventricle is the *aortic semilunar opening*, which contains the aortic semilunar valves. The interventricular septum is mostly muscular, but has a thin, tough membrane portion near its upper end.

Bodily Malfunctions

Errors during the heart's early development may produce openings ("holes in the heart") between the walls that separate the atria (atrial septal defects) or between the ventricles (ventricular septal defects). Small holes usually close on their own as the heart grows and cause no symptoms. Larger defects may have to be surgically closed to keep oxygenated and deoxygenated blood from mixing between the chambers and causing other lung and heart problems.

Wall-to-Wall Layers

The walls of each of the four heart chambers consist of three layers:

◆ **Epicardium.**   This is the outer layer of the heart. It is covered by a special tissue called mesothelium. Mesothelium is a simple squamous epithelial tissue.

◆ **Myocardium.**   "Myo" means "muscle." This middle layer consists of cardiac muscle cells. It's the part of the heart that contracts during heartbeats. The myocardium is thin in each atrium, but very thick (several millimeters) in the ventricles. The myocardium is three times thicker in the left ventricle than in the right ventricle because it must pump blood to the far reaches of the body.

◆ **Endocardium.**   The internal layer of the walls. It lines the chambers and valves of the heart and is exposed to blood. The inner surface of the endocardium is lined by a special simple squamous epithelium called *endothelium*, which also lines the inner surfaces of all blood vessels.

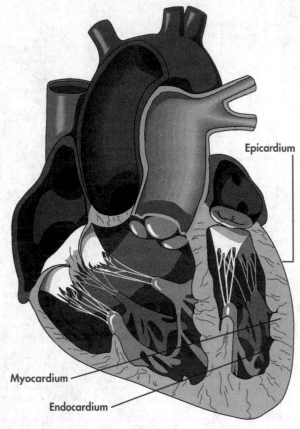

Layers of the heart.

Epicardium

Myocardium

Endocardium

Foot Notes

The ventricles have a different function than the atria. As a result, their walls, particularly the muscular part, are thicker than the walls of the atria.

Heart Valves: An Open-and-Shut Case

Heart valves are located at the openings between the four chambers. There are also valves between the chambers and the arteries leaving the heart. They control the direction the blood flows within and out of the heart.

The valves are normally smooth, delicate, supple flaps of tissue. Their outer margins are attached to the fibrous skeleton of the heart. The fibrous skeleton consists of dense connective tissue. It is located between the atria and the ventricles. This tissue encircles and firmly connects each of the four valve openings and serves as attachment for part of the heart muscle.

There are two varieties of valves:

◆ **Atrioventricular valves (AVs).** These are active valves that require muscular activity (contraction of papillary muscles) to close them. These valves are open during ventricular *diastole* (when the ventricles are relaxed) and closed during ventricular *systole* (when the ventricles are contracting). The right AV valve has three cusps and is also called the *tricuspid valve*. The left AV valve has two cusps and is also known as the *bicuspid* or *mitral valve*.

Body Language

The upper chambers (atria) and lower chambers (ventricles) of the heart each have periods of contraction **(systole)** and relaxation **(diastole)** during the heartbeat. The measurements recorded during a blood pressure reading reflects the highest systolic pressure when the ventricles contract and the remaining pressure (diastolic pressure) within the arteries when the ventricles relax.

◆ **Semilunar valves.** These valves are passive valves and do not require energy to close them. The *aortic semilunar valves* are located between the left ventricle and the aorta, whereas the pulmonary semilunar valves are located between the right ventricle and the pulmonary trunk. Each semilunar valve has three thin, floppy

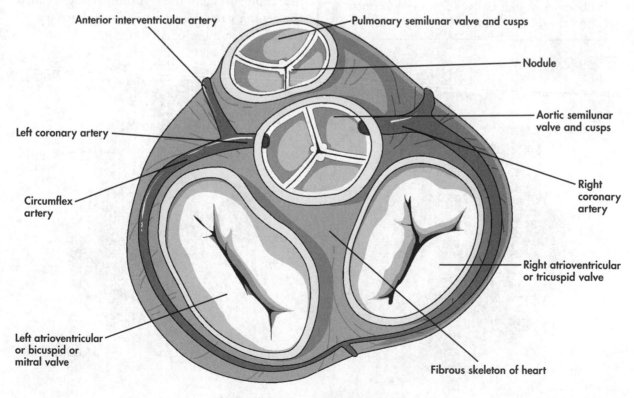

Fibrous skeleton and valves of the heart.

cusps that look like back pockets on a pair of jeans. At the center of each cusp is a small nodule of fibrous tissue. These cusps become back-filled with blood at the end of ventricular systole when the column of blood within the arteries loses energy and begins to fall back toward the ventricles. The nodules of the valves act like weights to ensure that the cusps close properly. These valves are closed during ventricular diastole.

Bodily Malfunctions

Valve disease commonly occurs years after someone has rheumatic fever. The inflammation that accompanies this disease may damage the valve cusps, causing them to become scarred and thickened. This interferes with blood flow. Valve damage may be picked up by a stethoscope on a routine physical exam by abnormal sounds of the valves as they open and close.

The Great Vessels and Beyond

The large veins and arteries that enter and leave the heart are called the great vessels of the heart. They include the inferior vena cava, the superior vena cava, the aorta, the pulmonary trunk, and the pulmonary veins.

Of these vessels, the venae cavae return oxygen-poor blood from the body to the right atrium of the heart. The pulmonary trunk and its immediate branches, the right and left pulmonary arteries, carry this venous blood from the right ventricle to the lungs, where it becomes oxygenated. Newly oxygenated blood is returned to the left atrium and is subsequently pumped from the left ventricle into the aorta to the body. The portion of the circulation between the heart and lungs is called the *pulmonary circulation*; the portion between the heart and aorta and the rest of the body is called the *systemic circulation*. I cover the pulmonary circulation in more detail in Chapter 9.

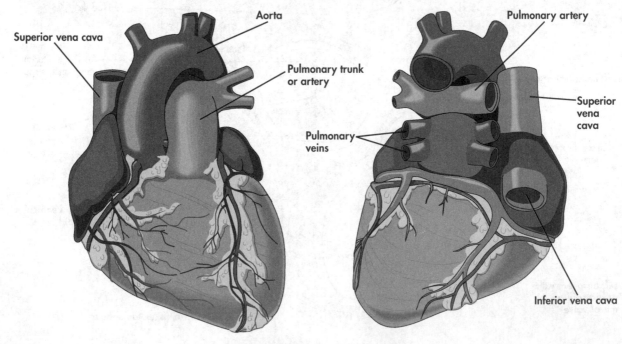

The great vessels of the heart.

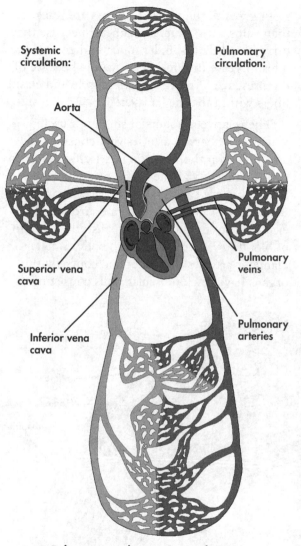

Systemic circulation:

Pulmonary circulation:

Aorta

Superior vena cava

Inferior vena cava

Pulmonary veins

Pulmonary arteries

Pulmonary and systemic circulation.

The Aorta and Major Arteries to the Body: Systemic Suppliers

The aorta receives oxygen-rich blood from the left ventricle and sends it to the body by way of its systemic branches. Arteries, anatomically speaking, carry blood away from the heart, whereas veins carry blood toward the heart. The major artery branches of the body are shown in the following figure. Each of these is covered in more detail in Part 3.

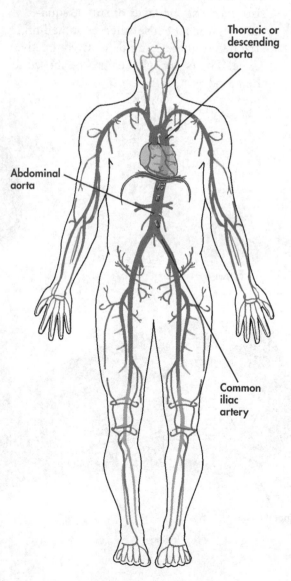

Thoracic or descending aorta

Abdominal aorta

Common iliac artery

Systemic arteries of the body.

Foot Notes

If the pulmonary artery carries venous blood, why is it not called the pulmonary vein? And, since pulmonary veins carry oxygenated blood, why are they called veins? Anatomically speaking, arteries carry blood away from the heart, whereas veins carry blood toward the heart. The names are given according to the direction of blood flow in reference to the heart and not according to the type of blood they carry. Got that?

The Anatomy of a Blood Vessel

Arteries and veins consist of three layers. These layers are similar to the layers of the heart's walls.

- ◆ **Adventitia.**   The outer layer, which is made up of loose connective tissue.

- ◆ **Media.**   The middle layer that consists of layers of smooth muscle cells and elastic fibers.

- ◆ **Intima.**   The internal lining layer consisting of a single layer of simple squamous epithelial cells, called endothelium, attached to underlying delicate connective tissue. The endothelial cells are exposed to the blood.

In general, the walls of arteries are thicker than veins with a similar diameter because they carry blood that is under much higher pressure. This increase in thickness is mainly because of the presence of more smooth muscle and elastic fibers within the media layer.

The relative amounts of smooth muscle and elastic fibers in arteries vary considerably depending on their function. Arteries closest to the heart have a greater concentration of elastic fibers (and are called *elastic arteries*) that enable them to withstand the high-pressure blood ejected from the heart and to push the column of blood toward the lungs and body. Smaller-diameter arteries that enter and course within organs have thick muscular walls consisting of

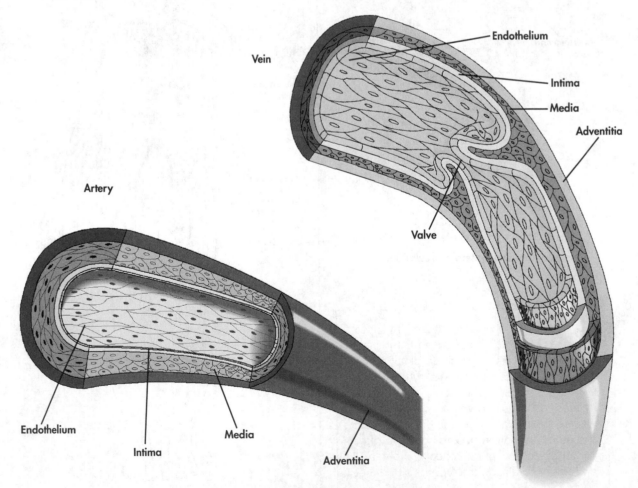

The anatomy of a typical artery and vein.

numerous layers of smooth muscle cells. These are called *muscular arteries*. As mentioned in Chapter 6, contraction of the smooth muscle in vessel walls is achieved by the sympathetic division of the autonomic nervous system. This contraction is important for the local regulation of blood flow and blood pressure.

Arteries that are less than 0.5 mm in diameter are called *arterioles*. They may have one or two layers of smooth muscle fibers within their walls. Arterioles usually connect to capillaries, which are discussed in the next section.

Bodily Malfunctions

Occasionally, the walls of an artery can weaken. The pressure against the weakened wall forms a bulge. This is called an aneurysm. Aneurysms can be life-threatening if they tear and cause internal bleeding.

The Very Capable Capillaries

Capillaries are the smallest of blood vessels. But they have one of the biggest jobs in the body. Capillaries are located in most tissue spaces of the body and connect small arteries to small veins. They are the body's narrowest and thinnest blood vessels, essentially consisting of tubes of endothelial cells. They are so small that a red blood cell (about 7.5 microns wide) has to squeeze its way along inside of them.

However, it is at the capillary-tissue level that exchange of oxygen, carbon dioxide, nutrients, and metabolic wastes takes place between the capillaries and the cells and tissues and vice versa. Water from the capillaries also leaks out between their endothelial cells to bathe nearby tissues. It has been estimated that the body contains about 60,000 miles of capillaries.

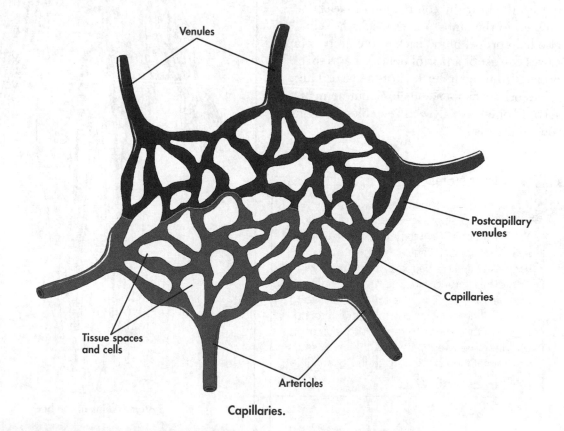

Capillaries.

Return Trip Home: The Vein Family

Veins are responsible for sending oxygen-poor blood back toward the heart. They begin as tiny *postcapillary venules* ("little veins") and join with increasingly larger-diameter venous channels to form visible vessels. These venules reabsorb most of the tissue fluid that originally leaked out the arterial end of the capillaries. Lymphatic capillaries reabsorb the remaining tissue fluid (which is discussed in the next chapter). If venous drainage from an area of the body is blocked or sluggish (for example, because of tight-fitting clothing or a blood clot in a large vein), the tissue becomes swollen. This condition is known as *edema.*

Veins freely interconnect other veins before assembling into main veins that run parallel to the major arteries. Veins direct their fluid cargo toward the heart by contraction of skeletal muscles in the limbs. Valves inside the veins also help propel blood back to the heart. These valves consist of a pair of delicate cusps that project from the inner wall of the vein. These valves passively open and close, but normally serve as one-way valves to keep blood moving toward the heart.

Foot Notes

"Arteries branch; veins join." Arteries branch into ever smaller and smaller branches until they become capillaries. Postcapillary venules join with other venules until they become veins. Veins join or are tributary to other veins until the largest veins in the body are formed. Think of postcapillary venules as small underground springs that join with creeks, and then rivers, and then the great Mississippi River (a major vein), until they all end up in the ocean (the heart).

You're So Vena Cava

Veins end by joining either the *inferior vena cava* or the *superior vena cava.* The superior vena cava returns blood from the brain, head, neck, upper limb, and thoracic wall. The inferior vena cava returns blood from the abdomen, pelvis, and lower limbs. Venous return from the entire body is then directed into the right atrium of the heart.

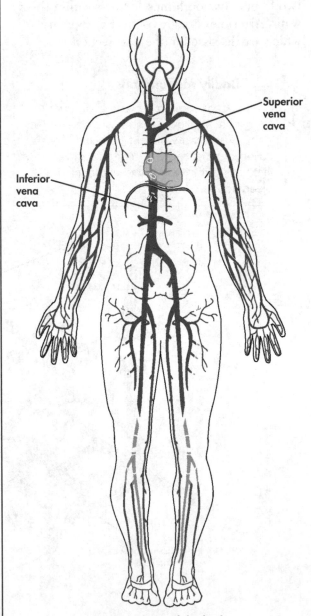

Systemic veins of the body.

The Portal Venous System: The Doors to Your Digestive Organs

The portal venous system begins in capillaries within the walls of the digestive tract from the stomach to the rectum. It ends in capillaries in the liver. The portal venous system delivers nutrients absorbed during digestion, including potential toxins, directly to the liver for their metabolism or detoxification. This blood leaves the liver via the *hepatic veins*, which empty into the inferior vena cava.

The portal vein is a large vein that is formed when the superior mesenteric vein joins the splenic vein. This is discussed in more detail in Chapter 18.

The Least You Need to Know

- There are two parts to the cardiovascular system: the heart, which acts as a pump, and the body-wide network of arteries and veins.
- The heart is a hollow, four-chambered organ with two atria and two ventricles.
- Blood flow within and out of the heart is controlled by sets of atrioventricular and semilunar valves.
- The pulmonary trunk and pulmonary veins are part of the pulmonary circulation that oxygenates the blood so that it can be pumped to organs, tissues, and cells of the body.
- Venous return from the digestive tract goes first to the liver via the portal system, whereas venous return from the rest of the body enters the superior or inferior vena cava before entering the right atrium of the heart.

In This Chapter

- ◆ Discover lymph and lymphatic capillaries
- ◆ The pattern that lymph drainage follows
- ◆ The lymphoid organs
- ◆ The basic function of the immune system

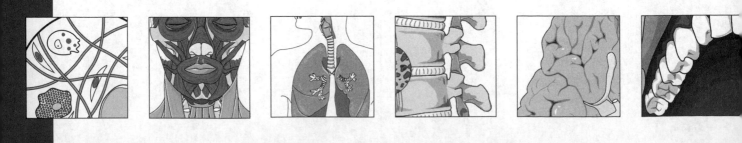

Chapter 8

Low Fluid Level: The Lymphatic System

The lymphatic system is a component of the circulatory system. It collects the excess tissue fluid leaked from blood capillaries in the body's tissue spaces and returns it to the venous system. This function is vital for maintaining normal blood volume within the circulation system.

The lymphatic system has other important functions, such as to absorb and transport fatty acids from the small intestine during digestion, to capture inhaled particulates from the air we breathe, and destroy bacteria or cancer cells that have invaded the body, and to produce immune cells as an integral part of the immune system.

In this chapter, you learn about these vital activities, the major lymphatic channels, and the lymphoid organs that are part of this system.

Where It All Begins: The Lymphatic Capillaries

In the last chapter, you learned that tissues and cells of the body are bathed in blood fluid that leaks out the arterial ends of blood capillaries. It is through this fluid that cells exchange their metabolic wastes and carbon dioxide for newly arriving nutrients and oxygen. Most of this tissue fluid is reabsorbed at the venous side of capillaries and returned to the heart by veins. A small but significant fraction of this tissue fluid, however, is not. Enter the lymphatic system.

The lymphatic system begins within most tissues of the body (except in the central nervous system, the eye, and bone) as dense networks of very thin-walled, closed-ended sacs called *lymphatic capillaries.* They are similar to blood capillaries in structure—that is, they are tubes of endothelial cells. Tissue fluid enters the lymphatic capillaries, which join and form progressively larger lymphatic vessels.

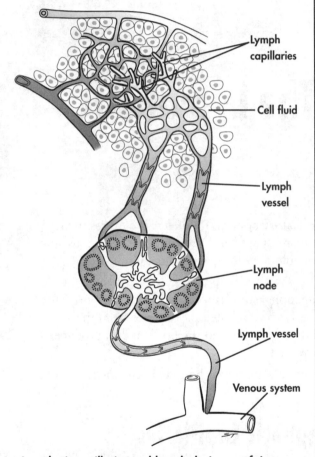

Lymphatic capillaries and lymph drainage of tissues.

There is no pump that drives lymph along its vessels. *Lymph* flow is dependent on a variety of movements and structures. Lymph vessels have numerous one-way valves that direct the fluid toward lymph nodes and the venous system. Lymph vessels have small amounts of smooth muscle within their walls that can contract to move lymph along (this activity is called *peristalsis*). Contraction of skeletal muscle during body movement, or pulsation of nearby arteries, compresses the walls of lymph vessels and assists lymph flow along these channels.

Body Language

Lymph is a translucent, straw-colored fluid that is similar to blood plasma, but with a lower concentration of proteins. The exception to this rule occurs in the small intestine following a meal. During digestion and the breakdown of fats, fatty acids are transported into the lymph vessels, making the lymph appear milky (the lymph here is called chyle).

The volume of lymph fluid that is recovered, filtered, and returned to the venous system each day is impressive—about two to four liters of fluid. This represents at least 73 percent of the total plasma volume in the body. This volume explains why blockage of lymph channels draining an area of the body causes local tissue swelling, called *lymphedema.*

In-line Filters: Lymph Nodes

At some point along its course, a lymph vessel meets and enters a lymph node. These are small filters consisting of a fine sponge-like meshwork of connective tissue fibers, containing *macrophages*, *lymphocytes*, and other immune cells. Lymph fluid percolates through this meshwork before leaving the node and entering another lymph vessel further up the line. Any foreign material, bacteria, or cancer cells present in the lymph are destroyed and removed from the lymph. Lymph nodes are very efficient and effective at neutralizing such material. More information about their function comes later in this chapter.

Body Language _____

Macrophages are cells that digest debris and invaders and stimulate other immune cells to respond to the threat. **Lymphocytes** destroy foreign cells. T-lymphocytes (or T-cells) destroy foreign or abnormal cells, and B-lymphocytes (or B-cells) produce antibodies during an immune response.

Where the Lymphatic Action Is

There is a predictable pattern of lymph drainage from tissues and organs of the body. Knowing these patterns is vitally important in determining the location of primary infections or tumors and proper care and treatment. For example, most (about 75 percent) of the lymph fluid from the breast first passes through the axillary lymph nodes (particularly the pectoral lymph nodes) located near the armpit (axilla). This explains why these nodes are frequently biopsied when breast cancer is suspected (and are removed when cancer is confirmed).

Lymph Trunks

Lymphatic vessels that carry lymph from regional groups of lymph nodes are called *lymph trunks*. These join to form larger lymph ducts. These unite to veins located behind each collarbone in the neck.

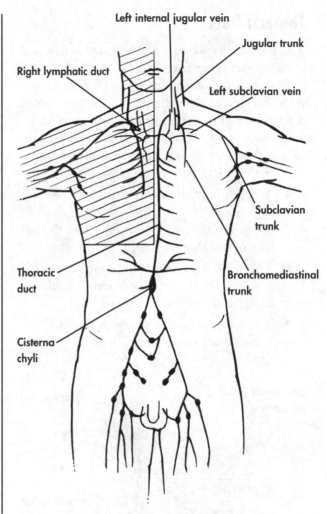

Lymph trunks and ducts and their terminations.

The *cisterna chyli* is a dilated sac located along the posterior wall of the abdomen. It receives lymph from the lower limbs, the pelvic organs, and the abdominal organs, including the digestive tract.

Thoracic Duct

The *thoracic duct* begins at the cisterna chyli and is the largest lymphatic vessel in the body. It courses upward on the surfaces of the thoracic vertebrae and ends at the base of the neck on the left side. There it enters the venous system at the junction between the left internal jugular vein and left subclavian vein. Near this termination, the thoracic duct receives the following:

◆ The left bronchomediastinal trunk, which returns lymph from the medial quadrants of the left breast, the left lung, and the left side of the thoracic cavity

◆ The left subclavian trunk, which returns lymph from the left upper limb, shoulder, axilla, and most of the breast

◆ The left jugular trunk, which returns lymph from the left half of the neck and head

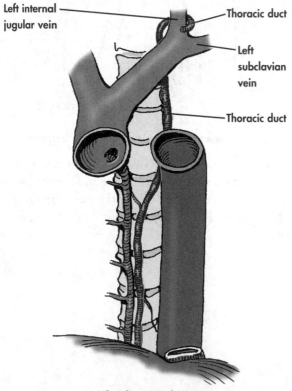

Left internal jugular vein

Thoracic duct

Left subclavian vein

Thoracic duct

The thoracic duct.

The thoracic duct transmits the majority of the lymph from the body, from the upper left half of the body and the whole body below the ribcage. The right lymphatic duct, on the other hand, has a much more limited portion of the body to drain. It is a short duct that receives similarly named trunks. It empties into the venous system at the junction between the right internal jugular vein and the right subclavian vein. It transmits lymph from the right upper half of the body.

Lymphoid Organs: The Pit Bulls of the Immune System

The body is continually under attack by external and internal invaders, such as bacteria, viruses, or the body's own cells. The immune system is the defense system that keeps these invaders at bay. It consists of an extremely sophisticated surveillance system of fixed and mobile immune cells that monitor and rapidly respond to assaults. This system utilizes an army of white blood cells, particularly lymphocytes and their hormone-like chemical signaling proteins, which travel throughout the body via the blood stream and lymph circulation.

All lymphocytes (B-lymphocytes and T-lymphocytes) are produced in the bone marrow. T-lymphocytes migrate to and mature within the thymus gland. The bone marrow and thymus are known as *central lymphoid organs.* Both T- and B-lymphocytes migrate via the blood stream and populate peripheral lymphoid organs such as the spleen, lymph nodes, tonsils, and collections of lymphoid cells in the mucosa of the small intestine (Peyer's patches), the appendix, and mucosal lining of the respiratory, urinary, and reproductive systems. Here, within these peripheral tissues, lymphocytes become competent, produce additional lymphocytes, and respond to antigens and other foreign materials.

The goal of the immune system, with its widespread collection of lymphoid tissues, is to destroy or neutralize microorganisms, abnormal cells, and foreign substances that invade the body. It does this efficiently in one of two ways:

♦ **Cell-mediated immunity.** This involves direct assault by activated T-lymphocytes to kill foreign cells. Autoimmune diseases, including rheumatoid arthritis, lupus erythematosus, and multiple sclerosis, are disorders caused when the body's immune system begins to attack its own normal tissues.

♦ **Humoral or antibody-mediated immunity.** This involves B-cells, following their activation by helper T-lymphocytes, and production of antibodies against antigens.

This arsenal of cells and the proteins produced by them work to keep us protected and healthy 24/7/365! Failures of this system lead to illness or disease.

Lymph Nodes

Lymph nodes are pea- or kidney-shaped structures that range in size from a few millimeters to a centimeter in diameter. They are located throughout the body, but are concentrated in the armpits, groin, sides of the neck, jaw, and back of the head. Numerous lymph nodes are also located in the thoracic, abdominal, and pelvic cavities.

Lymph nodes respond to antigens and foreign materials carried in the lymph fluid. On occasion, lymph nodes can become overwhelmed by bacteria or cancer cells. A lymph node that has become infected by bacteria is typically swollen and enlarged, tender to the touch, and soft. However, a lymph node that has been invaded by tumor cells is typically enlarged, painless to the touch, and hard. As such, lymph nodes serve as sentinels to tip us off regarding the presence of some previously unrecognized intruder.

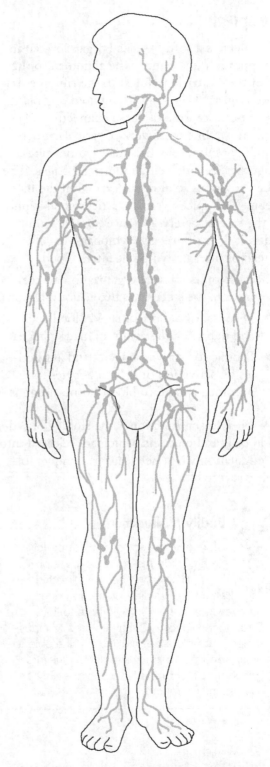

The lymphatic system and lymph node distribution in the body.

The Spleen

The spleen is a soft, fist-sized organ located in the upper left quadrant of the abdomen to the left of the stomach and next to the rib cage. It is the largest of the body's lymphoid organs. It has an outer capsule of dense connective tissue. Inside it is a fine meshwork of reticular fibers containing numerous B- and T-lymphocytes, antigen processing cells, and macrophages. The spleen responds to antigens carried in the blood stream. The spleen receives a rich blood supply via the splenic artery. However, unlike lymph nodes, it does not receive lymphatic vessels. Instead of filtering lymph, it filters blood.

Because of its internal organization and cell composition, the spleen has two major functions:

◆ Disposal of old or damaged blood cells and debris by the action of macrophages

◆ Production of lymphocytes and antibodies by activation of B-lymphocytes that become exposed to blood-borne antigens

A third, lesser, function is the storage of red blood cells and platelets, which are released into the circulation when needed, for example, following severe blood loss.

Bodily Malfunctions

Normally, the spleen cannot be felt from outside the body because of its small size and because it is hidden behind the left side of the rib cage. However, in certain infectious conditions, like mononucleosis, it may enlarge to many times its normal size, making it vulnerable to injury and possible life-threatening internal bleeding should it rupture spontaneously or due to minor trauma.

Injury to the spleen may require that it be removed. If the spleen is removed, other lymphoid organs of the body take over its functions.

The Thymus

The thymus is a primary lymphoid organ with two lobes, located in the upper chest behind the sternum. It actively grows in childhood and begins to regress during puberty. This organ is very important in establishing an individual's immune system.

Like the spleen, the thymus receives no lymphatic vessels. Its principal function is to produce fully competent T-lymphocytes from bone marrow cell precursors transported to the thymus via the blood stream. These cells then leave the thymus and enter the blood and lymph circulation, where they make up about 80 percent of the circulating lymphocytes. They are components of the cell-mediated immune response and are largely responsible for recognizing and destroying "non-self" cells.

Other lymphoid organs, such as the tonsils, Peyer's patches in the small intestine, the appendix, and lymphoid cells located within the lining of the respiratory, urinary, and reproductive systems, respond to antigens carried in tissue fluids associated with these systems. Thus, the body is "dotted" with immune tissues that monitor and quickly respond to unwanted intruders.

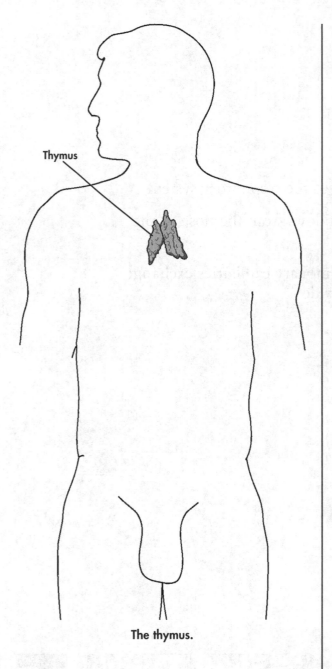

Thymus

The thymus.

The Least You Need to Know

◆ The lymphatic system plays a vital role in maintaining blood volume in the body by collecting excess tissue fluid that leaks out of capillaries.

◆ All lymph produced in the body passes through lymph nodes before it returns to the bloodstream.

◆ Lymphoid tissues are very efficient and effective at "cleansing" lymph, blood, or tissue fluids of bacteria or other foreign material.

◆ There are two classes of immune response: cell-mediated immunity and humoral or antibody-mediated immunity.

In This Chapter

◆ The parts and features of the respiratory system

◆ The pathway that air follows from the nose to the lungs

◆ How the lungs and pulmonary capillaries exchange oxygen and carbon dioxide

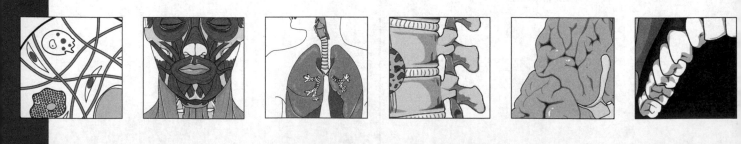

Chapter 9

You Take My Breath Away: The Respiratory System

The respiratory system consists of the lungs, where gasses (oxygen and carbon dioxide) are exchanged, and the passages that send air to and from the lungs. These passages include the nose, nasal cavities, mouth (oral cavity), pharynx, larynx, trachea, and bronchi. The primary function of the respiratory system is to supply blood with oxygen so that it can be delivered to the cells via the cardiovascular system.

In this chapter, you learn about the basic components of the respiratory system, the mechanics of breathing and gaseous exchange, and how this system performs this vital function. And I won't leave you breathless with too many details.

Look Out Below: The Upper Respiratory Tree

The lungs are the primary organs of the respiratory system. They are located in the thoracic cavity. But before air can get to the lungs, it must enter the upper portion of the respiratory tree.

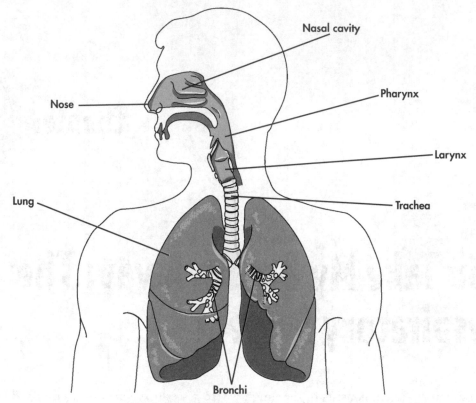

Overview of the respiratory system.

The respiratory system is made up of many interconnected parts. It extends from the head into the chest. The upper respiratory system is contained within the head and neck. Doctors and nurses (or TV ads) frequently talk about "upper respiratory infections," or URIs for short. Now you know what they're referring to!

Who Nose?: The Nasal Cavity

It all begins at the nose. The nose is the structure in the center of your face that you rest your glasses on. But it is much more than that. It consists mostly of cartilage and a thin pair of nasal bones. The cartilage and bones are covered by some thin, delicate muscles and skin. The nose has two openings called nostrils, or *external nares*. Each is lined with stratified squamous epithelium that contains coarse hairs for filtering out dust and debris from the air.

The nostrils are separated by the front portion of the *nasal septum*, which also consists of cartilage and some thin bone. Each nostril leads into larger chambers called the *nasal cavities,* which are located inside the head. There is a right and a left nasal cavity. The lateral wall of each nasal cavity has three scroll-like bones called *conchae* that project into the nasal cavity. The conchae increase the surface area over which air passes. The nasal septum, also referred to as the *medial wall of the nasal cavity,* is flat. Each cavity is covered by mucous membrane. The cavities are lined primarily by respiratory epithelium, which contains cilia and mucus-secreting goblet cells. The latter cells produce a blanket of sticky mucus that entraps dust and particulates, bacteria, and viruses in the air. The cilia beat in the direction of the throat, where the material is swallowed.

Beneath the mucous membrane is a rich bunch of capillaries that help warm the air. The nasal cavities condition the air we breathe by trapping and filtering out particulates, and humidifying and warming the air before it goes to the lungs.

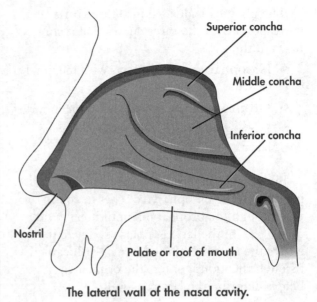

The lateral wall of the nasal cavity.

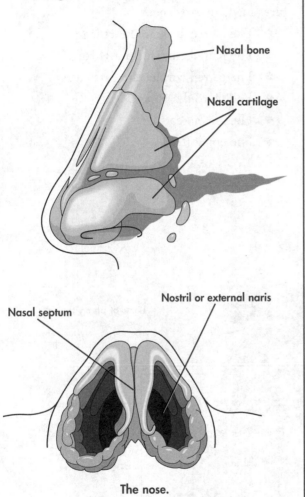

The nose.

Mouth (Oral Cavity): Take a Deep Breath

If you pinch your nose, you can still breathe through your mouth or oral cavity. From the mouth, air enters the throat or oropharynx on its way to the lungs. The mouth is generally described with the digestive system, which you will learn about in Chapter 10, but it often serves double duty as another passageway for conducting air.

Pharynx: The Throat

The pharynx is the next passageway that the air travels through on its way to the lungs. The pharynx is a funnel-shape muscular tube that attaches to the base of the skull and connects the nasal cavity to the esophagus in the lower neck. Its back wall lies in front of the cervical portion of the spine. Its front wall is open to three spaces in the head and neck: the nasal cavity, the mouth (or oral cavity), and the voice box (or larynx).

The pharynx is divided into three parts that correspond with the three spaces that it connects to the …

 ◆ **Nasopharynx.** Connects with the nasal cavity

 ◆ **Oropharynx.** Connects with the mouth

 ◆ **Laryngopharynx.** Connects with the larynx

Breathed-in air from the nasal cavities passes into the nasopharynx. Then it continues through the oropharynx until it gets to the upper part of the laryngopharynx. There it enters the opening of the larynx. Air breathed through the mouth enters the oropharynx before entering the larynx as well.

Larynx: The Voice Box

The larynx (which is commonly called the voice box) is located in the midline of the neck between the pharynx and the trachea. The framework of the larynx is made of nine structures made of cartilage:

 ◆ The paired arytenoid cartilages
 ◆ The paired cuneiform cartilages
 ◆ The paired corniculate cartilages
 ◆ The thyroid cartilage
 ◆ The cricoid cartilage
 ◆ The epiglottis

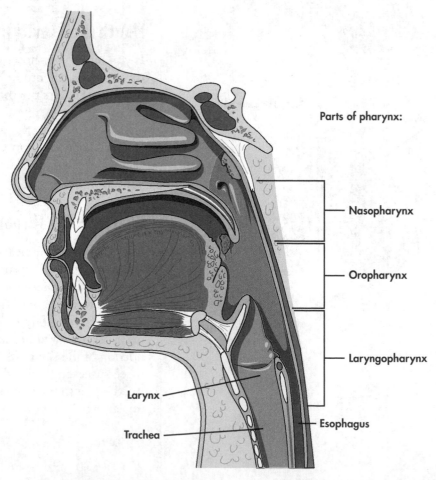

Parts of pharynx:

Nasopharynx

Oropharynx

Laryngopharynx

Larynx

Esophagus

Trachea

The pharynx.

The largest of these is the thyroid cartilage. It looks like a shield. The thyroid cartilage sticks out on the front of the neck beneath the skin of adult males. This thing that sticks out is called the laryngeal prominence, better known as the Adam's apple. These cartilages are connected to one another by ligaments, membranes, and skeletal muscles. All of them are covered by a moist mucous membrane.

The larynx has two pairs of synovial joints. One pair is located between the thyroid and cricoid cartilages. The other pair occurs between the cricoid and arytenoid cartilages. Muscles move these joints to alter the size of the opening of the airway and the tension on the vocal cords.

The entrance to the inside of the larynx is located behind the large spatula-shape epiglottis and between a pair of membrane folds that connect the sides of the epiglottis to the arytenoid cartilages. The lower edge of this membrane fold forms a pair of bulky folds called the *vestibular folds* that project into the interior of the larynx. These are also called the *false vocal cords* because they have no role in sound production. Within the larynx, below the vestibular folds, is a pair of elastic membranes containing the vocal cords, which form the true vocal cords. The vocal cords connect from the arytenoid cartilages to the inner surface of the thyroid cartilage. The space between these vocal cords is the path that air takes to and from the lungs. Muscles change the size of this opening. It is open widest during vigorous exercise, and open the least when a person is speaking.

The vocal cords are forcefully closed during swallowing to keep unwanted food materials or fluids from passing into the trachea and lungs.

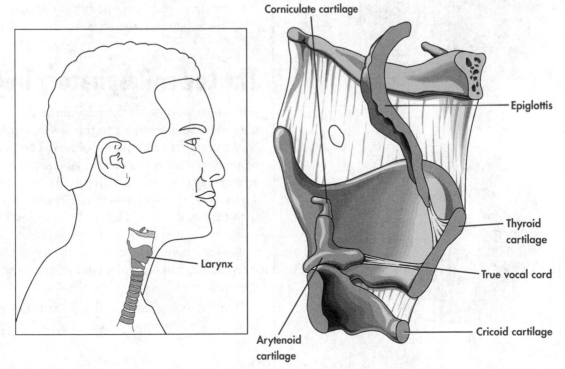

The larynx.

During swallowing, the epiglottis is also folded over the entrance to the larynx to protect the airway.

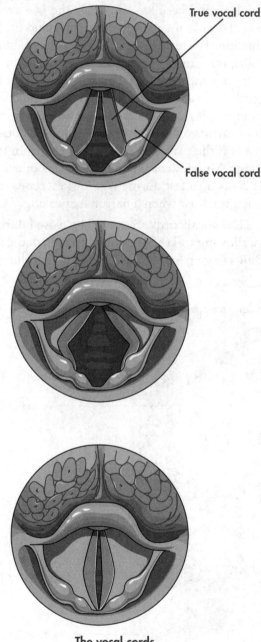

True vocal cord

False vocal cord

The vocal cords.

The lining of the larynx around the vocal cords is richly supplied by sensory nerves. It is extremely sensitive to materials "going down the wrong pipe." When this happens, it causes an immediate, explosive cough reflex that, hopefully, drives out the misdirected material.

The larynx has three important functions:

◆ **Protection of the airway.** This is its most important function. The larynx does this by bringing the true vocal cords together to form a tight seal.

◆ **Phonation, or voice production.** Sound is produced when a pressurized column of air from the lungs is forced through the narrow slit between the vocal cords, causing them to vibrate. This column of air is acted upon above by the walls of the pharynx, the soft palate, the tongue, the teeth, and the lips to produce intelligible speech.

◆ **Respiration.** This is the simple transfer of air to and from the lungs through the space between the vocal cords.

The Lower Respiratory Tree

The upper portion of the respiratory tree conditions and sends air to the lower respiratory tree, where respiration occurs. The lower respiratory tree is the functional end of the respiratory system. Tiny air sacs in the lungs are surrounded by capillaries that exchange oxygen with carbon dioxide. This exchange allows the cardiovascular system to deliver oxygen to all cells of the body. The cells use the oxygen to convert nutrients into the energy they need for their activities.

The following sections detail the major parts of the lower respiratory tree.

Trachea and Bronchi: The Trunk and Branches

The trachea (or windpipe) is connected to the lower end of the larynx. It passes from the neck into the upper part of the thorax before dividing into right and left bronchi. The trachea is about five or six inches long and one inch wide in an adult. It lies in front of the esophagus. The trachea consists of a series of 16 to 20 C-shape cartilage rings. These rings' backward-projecting free ends are bridged by smooth muscle and connective tissue. The cartilages keep the airway open and prevent the trachea from collapsing during breathing.

The trachea divides into right and left primary bronchi that are directed toward the right and left lungs. Each bronchus divides into secondary (or *lobar*) bronchi that enter the lobes of each lung. The right lung has three lobes, whereas the left lung has two lobes. Within each lobe, the secondary bronchus divides into tertiary (or segmental) bronchi. These bronchi conduct air to specific parts of each lobe. Hereafter, these bronchi progressively divide into smaller and smaller bronchi, like the branches on a tree.

The structure of the bronchi is similar to the trachea. But when the bronchi are inside the lung, their circumference contains islands,

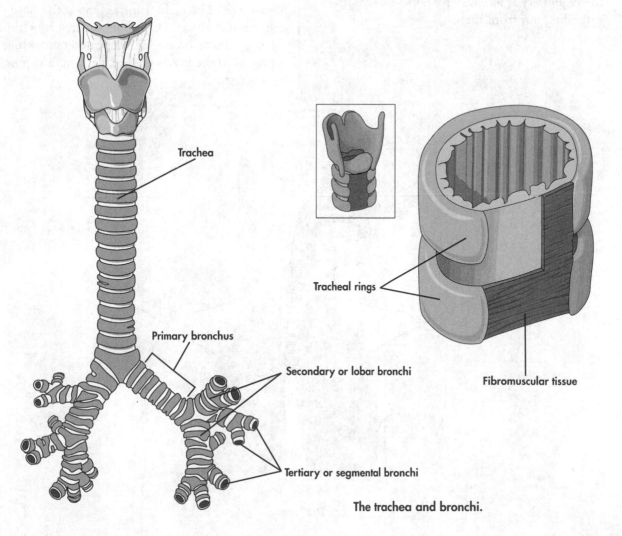

Trachea

Primary bronchus

Secondary or lobar bronchi

Tertiary or segmental bronchi

Tracheal rings

Fibromuscular tissue

The trachea and bronchi.

or plates of cartilage. Internal to these islands are smooth muscle and mucous membrane. The lining of these passageways is respiratory epithelium. Ciliary action along the lining of the bronchi and trachea beat entrapped dust or bacteria upward toward the pharynx, where it is swallowed.

Continued branching produces passageways that are approximately one millimeter in diameter. These passages' walls lack cartilage and consist of smooth muscle that is lined by a simple epithelial covering that doesn't have glands or cilia. These passageways are called *bronchioles*. They also branch, become smaller in diameter, and get thinner walls. The smallest of these, the *respiratory bronchioles*, have sac-like spaces extending out from them.

Bodily Malfunctions

Asthma is a chronic disease of the airways that causes their linings to be swollen and very sensitive. When these airways are exposed to irritants or allergens, they constrict and make it difficult to breathe (also known as an asthma attack). Asthma cannot be cured, but it can be controlled with medications.

Lungs: The Leaves of the Respiratory Tree

Think of the respiratory system as an upside-down tree. The main trunk begins at the nose and nasal cavity and continues to the bronchi before it starts branching. Repeated branchings bring us to the leaves of the tree, where *respiration* occurs.

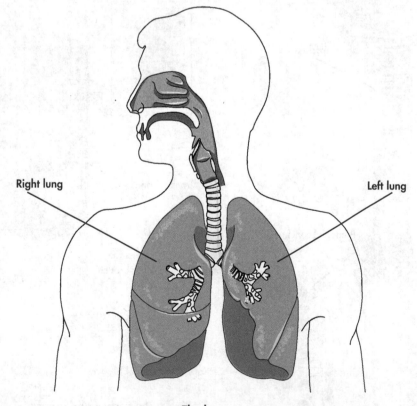

The lungs.

Body Language _____

Respiration is the process of breathing, in which carbon dioxide is taken from blood cells and replaced with oxygen.

Foot Notes _____

Premature babies are often born without enough surfactant in their lungs, making it very difficult for them to breathe. Mothers who are about to deliver prematurely can be given a hormone shot that increases the baby's production of surfactant and gives them a better chance of survival.

The respiratory bronchioles at the outermost reaches of the respiratory tree connect to *alveolar ducts*. These ducts end in grape-like clusters of sacs, called alveoli. *Alveoli* are the structural and functional unit of respiration. There are approximately 300 million of these air spaces in each lung! The wall of each alveolus is very thin and consists of a single layer of squamous epithelial cells. Between the walls of adjacent alveoli are pulmonary capillaries. These capillaries are where the exchange of oxygen and carbon dioxide occurs.

This very efficient process is called *respiration*. Interspersed within the walls of alveoli are special cells that produce a detergent-like substance called *surfactant*. Surfactant helps to reduce the surface tension within the fluid that moistens the flimsy alveoli, keeping them open during respiration and not collapsing on each breath.

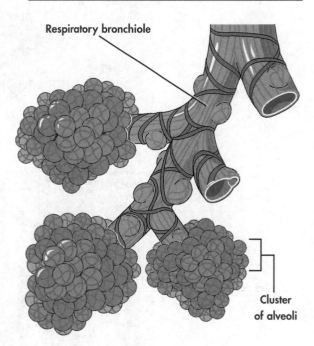

Respiratory bronchiole

Cluster of alveoli

Respiratory bronchiole and alveoli.

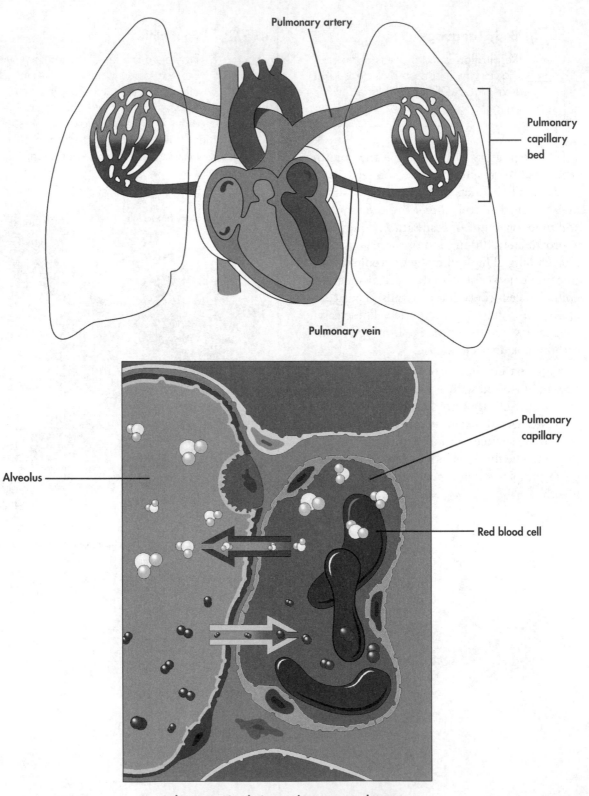

Pulmonary circulation and gaseous exchange.

In with the New, Out with the Old: Gaseous Exchange

The primary function of the respiratory system is to supply blood with oxygen so that it can be delivered to the cells of the body via the cardiovascular system. At this same time, the respiratory system is getting rid of the toxic gas, carbon dioxide. This gaseous exchange is accomplished by breathing, which involves contraction of muscles that allow air to enter and leave the lungs. The mechanics of this process of gaseous exchange are discussed in detail in Chapter 17.

The Least You Need to Know

◆ The respiratory system collaborates with the cardiovascular system to provide oxygen to all cells of the body.

◆ The respiratory system consists of a series of air passageways that clean and condition the air before it gets to the alveoli of the lungs.

◆ Gaseous exchange of oxygen and carbon dioxide occurs only within alveoli, between these air spaces and the pulmonary capillaries.

In This Chapter

◆ The parts of the digestive system

◆ The unique features of each part of the digestive system and how each part works

◆ The accessory organs of digestion and how they contribute to the digestive process

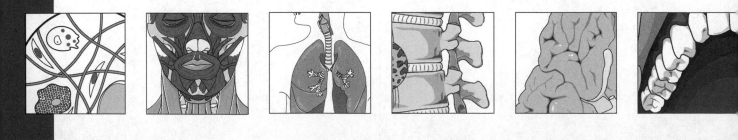

Chapter 10

The Long, Winding Food Tube: The Digestive System

Do I live to eat? Or, do I eat to live? Sometimes I have different answers to those questions. Either way, the digestive system is engaged in this activity of joy or necessity.

The digestive system extends from the mouth to the anus, a distance of about 30 feet. Each section along the way has a different but unique purpose that supports its mission: to disassemble foods into their smallest units (amino acids, simple sugars, and fatty acids) so that vitamins, minerals, and nutrients can be absorbed into the blood stream and dispersed to the body's cells for their energy and growth needs. This system also prepares undigestible materials for elimination from the body as feces. Digestive function is enhanced and made more efficient by several accessory organs of digestion, such as the salivary glands, the pancreas, and the liver and gallbladder.

Eating is, indeed, one of the great pleasures of life. This chapter takes you step-by-step (or perhaps, better, curve-by-curve) down the digestive tract and highlights the functional features of this system.

Chew on This! The Mouth

Solid food that a human eats needs to be physically reduced in size and consistency so that it can be safely swallowed. This occurs in the mouth, where the teeth shred, crush, and break down the food. This is made possible by muscles of mastication (chewing) that move the lower jaw and teeth against the upper jaw and teeth.

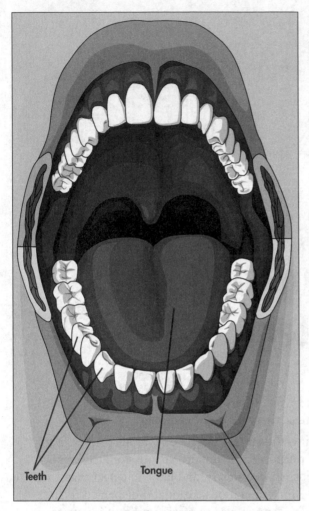

The mouth.

The Tongue

The tongue, in addition to containing taste buds that enhance a person's enjoyment of food, is vitally important in this endeavor. The tongue is a muscular, mobile organ that acts like a tumbler to move, churn, and reposition food between the teeth and against the *palate* (the roof of the mouth). In fact, paralysis of the tongue is devastating to a person's ability to break down food for swallowing.

Bodily Malfunctions

A person's sense of smell, or the lack thereof, plays a big role in their sense of taste. The nerve for smell (the olfactory nerve) is located high within the nasal cavity. When a person has a cold or severe allergies, the lining of the nasal cavity swells and blocks air from getting to this nerve to stimulate it. When this happens, a person loses part of the apparatus for tasting.

Salivary Glands

The breakdown of food is made easier by infusions of saliva from nearby salivary glands. Saliva moistens the food and makes it easier to move about in the mouth. There are three pairs of major salivary glands:

- **Parotid.** The parotid glands (*para* means "next to" and *otis* means "ear") are the largest of the salivary glands. They are located on the sides of the face just in front of the ear. Each parotid gland's duct leaves the gland, crosses the central region of the face, and passes through the wall of the cheek to open in the mouth via a small, fleshy projection. The saliva of the parotid gland is clear and watery.

- **Submandibular.** The submandibular glands are located along the inside of the lower jaw.

- **Sublingual.** The sublingual glands are located in the floor of the mouth beneath the tongue. The ducts of both the submandibular and sublingual glands open along the floor of the mouth. Their saliva is stickier.

Saliva consists mostly of water and salts and contains the enzyme *amylase*, which helps break down starches. It also contains an enzyme called *lysozyme*, which kills certain bacteria present on foods.

Saliva is important not only for preparing food for swallowing, but also for the health of the teeth by rinsing away bacteria and acids that may harm the enamel. Finally, saliva contains bicarbonate ion and has a basic pH level that allows it to protect the lining of the lower esophagus from stomach acid that may reflux or regurgitate into the lower end of the esophagus during digestion.

Down the Hatch: The Pharynx and Esophagus

As briefly described in Chapter 9, the *pharynx* is a funnel-shaped muscular tube whose front wall is missing. This allows it to communicate freely with the nasal cavity, the oral cavity, and the entrance to the larynx. It's the upper part of both the respiratory and the digestive systems. Its wall consists of an incomplete outer layer of circularly arranged skeletal muscle and an incomplete inner layer of longitudinal muscle. These muscles help elevate and sequentially contract to move swallowed food downward. The pharynx is connected to the esophagus in the lower neck.

The *esophagus* is a 10-inch-long muscular tube that descends from the neck into the thoracic cavity in front of the vertebral column. It simply transmits foodstuffs from the pharynx to the stomach. Its internal passage (called the *lumen*) is normally closed. Just like the pharynx, the wall of the esophagus consists of two layers of muscle: an inner layer of circularly arranged muscle fibers and an outer layer of longitudinally oriented muscle fibers. The upper end of the esophagus is composed entirely of skeletal muscle, which gradually transitions to smooth muscle. The lower end of the esophagus is composed entirely of smooth muscle, as is the remainder of the digestive tract. The muscles of the esophagus move the food along its length with contractions.

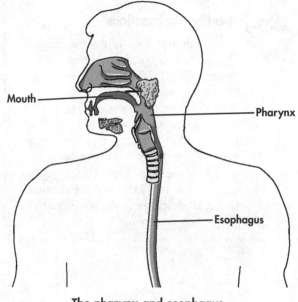

The pharynx and esophagus.

 Foot Notes

Although a person voluntarily controls his or her chewing, the act of swallowing is actually an automatic reflex. When the chewed food reaches the oropharynx, the swallow reflex takes over. Swallowing is a complex neuromuscular activity that involves structures in the pharynx, larynx, and esophagus.

Both the pharynx and the esophagus are lined by stratified squamous epithelium. This lining is resistant to abrasion by coarse food, such as corn chips. Contained within these linings are mucus-secreting glands. These secretions lubricate the surface for smooth passage of food. The mucus also serves as a protective barrier against the effects of stomach acid that may flow backward into the lower esophagus. The esophagus passes through the diaphragm before it connects to the stomach.

Bodily Malfunctions

Although the lining of the esophagus is resistant to a variety of food textures and small amounts of stomach acid, it is sensitive to backward flow of acid material from the stomach. The painful burning sensation that people feel when this happens is called "heartburn," even though the heart isn't involved in this condition whatsoever. When acid flows back into the esophagus on a regular basis, the condition is called *gastroesophageal reflux disease (GERD)*, or just *acid reflux*.

Pass Me an Antacid: The Stomach

The stomach is an enlarged portion of the digestive tract where digestion begins in earnest. It is a muscular reservoir with valves (called *sphincters*) at each end, which control passage of digested materials. Here, semisolid foods are mixed with gastric juices. These juices containing hydrochloric acid and activated enzymes that break down carbohydrates, proteins, and fats. Muscular activity "churns" and mixes the contents to reduce it to a viscous, semifluid mass called *chyme*. At the appropriate times, the pyloric sphincter relaxes, allowing small portions of chyme to enter the next segment of the gastrointestinal tract, the small intestine.

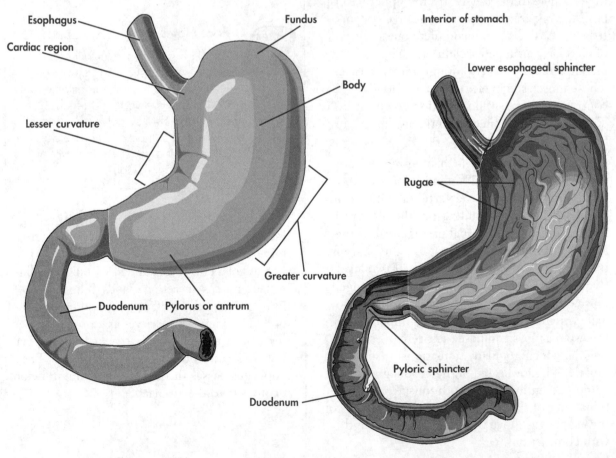

The stomach.

Roadmap of the Stomach

The stomach is described as a J-shaped organ. It is located in the upper-left region of the abdominal cavity, just below the diaphragm. To its left is the spleen, to its right is the liver, and behind it is the pancreas. The stomach has a pair of curvatures (greater and lesser) and four parts:

- **Greater curvature.** A convex border that faces the spleen.
- **Lesser curvature.** A concave border that faces the liver.
- **Cardiac region.** Where the stomach connects to the esophagus.
- **Fundus.** The part of the stomach to the left of the cardiac region where the stomach forms a dome.
- **Body.** The major part of the stomach that narrows as a funnel-shaped portion called the *pylorus* or *antrum*. This end attaches to the small intestine.

Ridges, Pits, Glands

When the stomach is empty, its lining has several longitudinal folds or ridges called *rugae*. These flatten out as the stomach fills during mealtime. This lining is covered by simple columnar epithelial cells. At the bases of these infoldings, called *gastric pits*, are the openings to tube-shaped *gastric glands*. These glands produce the gastric juices and hormones that are important for the activities of the stomach.

Body Language _____

Gastric or **gastro-** refers to anything in or near the stomach.

Layers, Muscles, and Sphincters

The stomach is unique among organs because it has three layers of smooth muscle in its wall:

- An inner oblique layer
- A middle circular layer
- An outer layer of longitudinal muscle

These muscles effectively churn stomach contents during digestion and wring them out as they are headed to the small intestine.

At both ends of the stomach are sphincters, or valves, that control entrance and exit of food material. They are of two types. At its upper end is the *lower esophageal sphincter*. This type of sphincter is known as a *physiological sphincter*. It lacks a distinctively thickened circular layer of smooth muscle of the found in the *anatomic sphincter*, but nonetheless it maintains closure until the food arrives from the pharynx. Then it relaxes to allow the food to pass into the stomach. At its lower end is the *pyloric sphincter*. It's considered a true anatomic sphincter because it has a thick layer of circularly arranged smooth muscle fibers. It is a strong sphincter and opens intermittently to allow passage of a small amount of chyme into the small intestine.

Body Language _____

A **physiological sphincter** is a valve that lacks a distinctive ring of smooth muscle whereas an **anatomical sphincter** is one that is made of a thick ring of smooth muscle. Both types are closed until autonomic nerves cause them to relax and open when needed.

Pylorus means "gatekeeper," which is an appropriate name for this region of the stomach. Here the **pyloric sphincter** controls passage of gastric contents into the small intestine.

Fully Absorbed: The Small Intestine

The small intestine is where the processes of digestion are completed and nutrients are absorbed into the blood stream and lymph system. This part of the digestive tract is about 20 feet long. It consists of three parts: the duodenum, the jejunum, and the ileum. It extends from the stomach to the first part of the large intestine.

The lining of the small intestine (called the *mucosa*) is highly amplified for maximum absorption efficiency. First, it consists of millions of microscopic finger-like extensions, called *villi*, that project into the passageway. Second, it contains numerous permanent circular folds, called *plicae circulares*. And, third, the simple columnar absorptive cells lining the mucous membrane have hundreds of microvilli on their surfaces. These are cylindrical extensions of cell membrane that make the cells look like the end of a brush. Collectively, these three features increase the surface area for absorption approximately 600-fold.

The wall of the small intestine contains two complete layers of smooth muscle; an inner circular layer and an outer longitudinal layer.

The Duodenum

The first and shortest part of the small intestine is the *duodenum*. It is "C"-shaped and holds the head and neck of the pancreas in its concavity. Through the concave side of this wall it receives the ends of the common bile duct and the main pancreatic duct. These ducts send their secretions of bile and pancreatic juices, respectively, into the duodenum. Digestive breakdown of the chyme continues here. The duodenum contracts to move the chyme toward the jejunum.

Body Language

Duodenum means "12 fingers," which refers to the fact that it's as long as 12 fingers are wide (or about 10 to 12 inches).

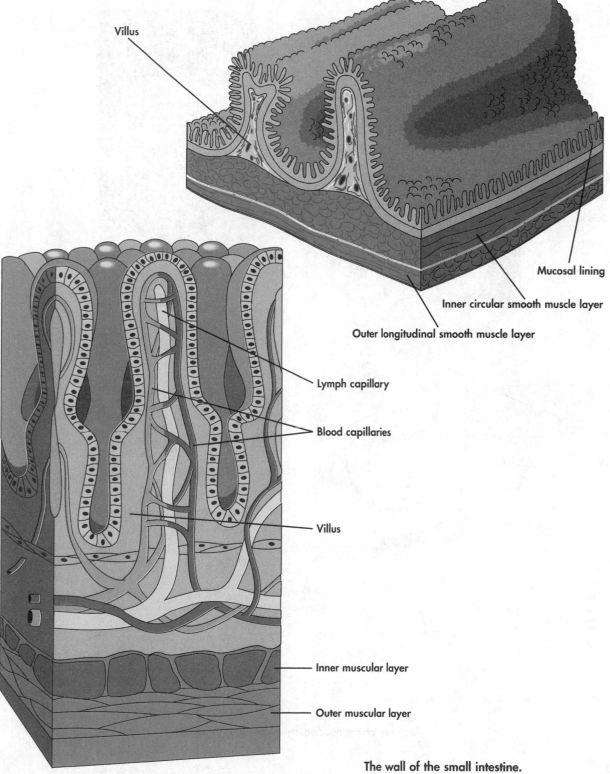

Villus

Mucosal lining

Inner circular smooth muscle layer

Outer longitudinal smooth muscle layer

Lymph capillary

Blood capillaries

Villus

Inner muscular layer

Outer muscular layer

The wall of the small intestine.

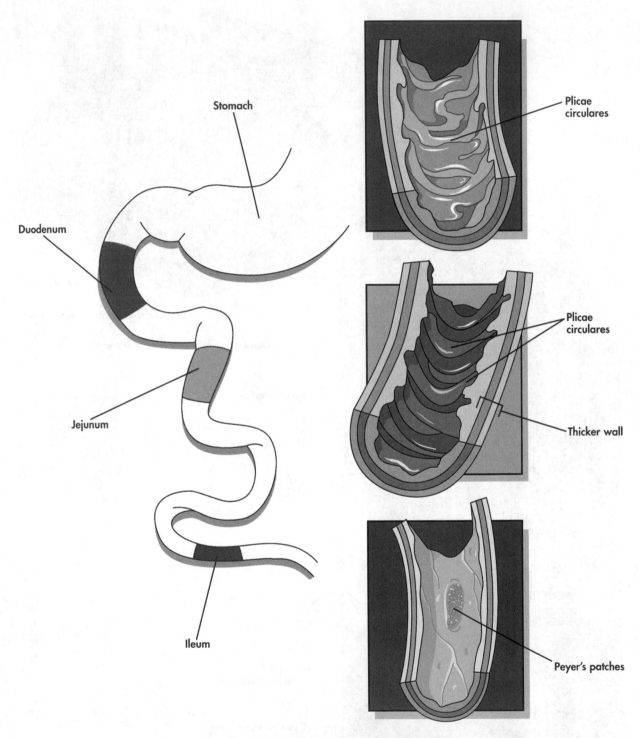

Features of the duodenum, jejunum, and ileum.

The Jejunum

The *jejunum* is the next segment of the small intestine. It is about 8 feet long. Its wall is thicker than that of the next segment, the ileum, and has the greatest density of plicae circulares within it. Bound to its microvilli are enzymes which help finish off carbohydrate and protein digestion. The sugars and amino acids this digestion produces are readily absorbed by the epithelial cells. Then they are transported to nearby blood capillaries associated with the portal venous system. Nutrients from digestion are carried within this system to the liver first before being sent on to the heart.

In addition, within the jejunum, the chyme mixes with enzymes from the pancreas and bile salts from the liver and gallbladder. Fats are broken down and absorbed by the epithelial cells. Ultimately, the fats absorbed by the small intestine along with other lymph fluid are transported to a lymph sac called the *cisterna chyli*, from which the thoracic duct begins. The thoracic duct delivers this intestinal lymph to the venous system.

The Ileum

The ileum is about 11 feet long. It has fewer plicae than the jejunum does. Its lining contains numerous collections of lymphoid tissue, called Peyer's patches. These collections of immune cells are part of the surveillance component of the immune system. They monitor antigens being passed along within the chyme. Both the jejunum and ileum are coiled and suspended by a membrane and fill much of the abdominal cavity.

The ileum terminates by entering the cecum of the large intestine. Here, the *ileocecal valve* regulates passage of the chyme into the large intestine.

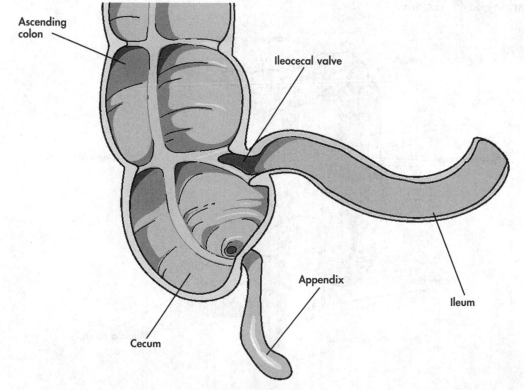

The ileocecal junction.

The End Is in Sight! The Large Intestine

Intestinal contents from the ileum enter the large intestine as a thin, watery mixture. The large intestine has three functions:

- ◆ To reabsorb water and minerals from the intestinal contents
- ◆ To form the fecal mass in preparation for its elimination from the body during a bowel movement
- ◆ To produce mucus to lubricate the intestinal surface for this journey

In other words, that which the body can't use is destined for the toilet bowl!

The large intestine is approximately five feet long. It consists of the cecum, appendix, ascending colon, transverse colon, descending colon, sigmoid colon, rectum, and the anal canal. The large intestine ends at the anus.

The mucous membrane lining the large intestine is relatively smooth, lacks villi, and contains only a few semilunar mucosal folds. Its simple columnar epithelial lining consists mainly of absorptive cells and numerous mucus-secreting goblet cells. These latter cells lubricate the intestinal surface for transit of the fecal mass.

Thanks, We Needed That! Accessory Organs of Digestion

The accessory organs of digestion provide vital and essential substances that allow the body to get all that it can out of the food people eat and keep their intestinal tracts healthy. Enzyme deficiencies, malnutrition, or decreased salivary gland secretion may cause digestive disturbances.

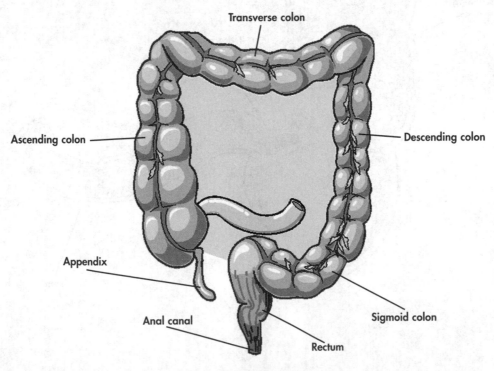

The large intestine.

These accessory organs of digestion include the salivary glands (see above), the liver, the gallbladder, and the pancreas.

The Pancreas

The pancreas is an elongated gland located along the back wall of the upper portion of the abdominal cavity. It consists of a head and neck that rest within the concavity of the duodenum, a body that is located behind the stomach and the transverse colon, and a tail that is directed toward the hilum of the spleen.

The pancreas is called a mixed gland because it has both an exocrine portion and an endocrine portion:

◆ **Exocrine pancreas.** Produces a watery solution that is rich in bicarbonate ion and about 15 enzymes that break down carbohydrates, fats, proteins, and nucleic acids. This fluid is transported to the duodenum through the pancreas's main and accessory pancreatic ducts. The enzymes in this fluid are inactive until they come into contact with the chyme. Then they are converted to their active forms. This keeps the pancreas from digesting itself.

◆ **Endocrine pancreas.** Secretes insulin and glucagon to help metabolize carbohydrates. These secretions directly enter the bloodstream.

The Liver

The liver is located in the right upper quadrant of the abdominal cavity, beneath the respiratory diaphragm. It is the largest internal organ in the body. Not surprisingly, it's important and has many functions:

◆ Processing and storing nutrients absorbed in the digestive tract

◆ Neutralizing and eliminating toxic substances

◆ Producing bile and important blood plasma proteins

◆ Serving as a reservoir for certain minerals and fat-soluble vitamins

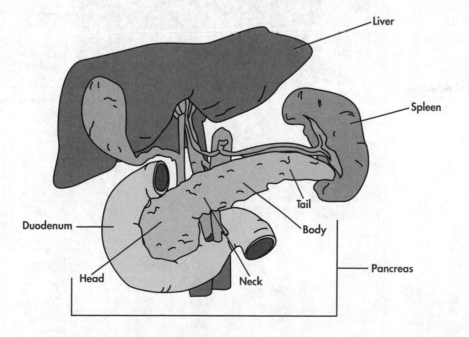

The pancreas.

The liver is unique in that it gets its blood supply from two different sources. About 80 percent of the blood to the liver enters via the portal vein, which receives nutrient-rich (except for lipids) but oxygen-poor blood from the abdominal organs. The remaining 20 percent is provided by the *hepatic artery*, which carries oxygen to support the activities of the *hepatocytes* (liver cells). Complex lipids and certain vitamins, especially vitamin A, arrive at the liver by this source.

The liver stores carbohydrates and lipids as glycogen and triglycerides, respectively, immediately following digestion. It supplies the body with energy sources between meals. Hepatocytes can also convert lipids and amino acids into glucose. And the liver is the main site of production of urea that is eventually excreted by the kidneys as urine.

The liver is an accessory organ of the digestive system because it produces bile. Liver cells continuously produce bile and excrete it into tiny channels called *bile canaliculi*. These merge to form *ductules* and eventually *hepatic ducts*, which leave the right and left lobes of the liver. The right and left hepatic ducts form the common hepatic duct, which joins the cystic duct

Body Language

Hepatic or **hepato-** refers to anything of or related to the liver.

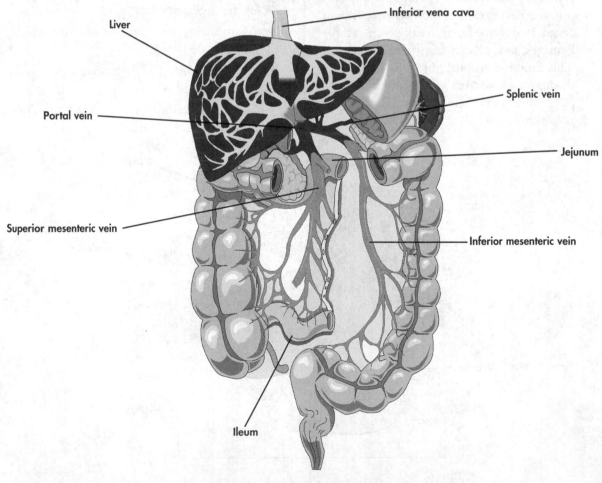

The portal venous system.

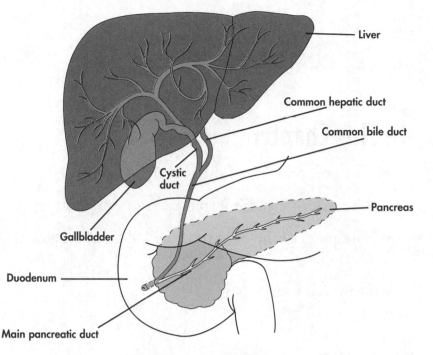

The liver and gallbladder.

from the gallbladder to form the *common bile duct*. The common bile duct courses downward toward the duodenum. It passes through a portion of the pancreas where it unites with the main pancreatic duct to pass through the duodenal wall.

Bile consists of water, ions, bile salts, cholesterol, and bilirubin. Bile acids are important in digestion of lipids by helping emulsify the fats and assisting in their breakdown.

The Gallbladder

The gallbladder is a pear-shaped structure that is attached to the undersurface of the liver. While bile is continuously produced by the liver, it is stored in the gallbladder. It holds about 30 to 50 milliliters of bile. The gallbladder can concentrate the bile several fold by absorbing its water. During digestion, cells in the duodenum release a hormone in response to the presence of fats in the chyme that causes the gallbladder to contract and the sphincter surrounding the opening of the common bile duct and main pancreatic duct to relax. This allows transport of their secretions into the duodenum to help with digestion.

The Least You Need to Know

◆ The digestive system physically and chemically disassembles the food a person eats into basic units that are absorbed into the circulatory system for use by the body.

◆ The digestive system extends from the mouth to the anus and each part of this pathway has unique functions.

◆ The digestive system is also responsible for eliminating food materials from the body that it cannot digest.

◆ The process of digestion is complex and requires substances produced by accessory organs to ensure proper and complete digestion of food.

In This Chapter

◆ The components of the urinary system

◆ The external and internal parts of the kidney

◆ The parts of the nephron, the kidney's functional unit

◆ How urine is made

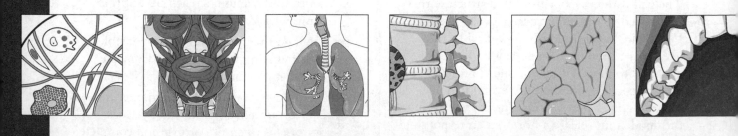

Making Water: The Urinary System

Just as the lymphatic system cleans microorganisms, abnormal or damaged cells, and other foreign substances from blood and lymph fluid, the urinary system rids the blood of waste products of cell metabolism and excess water. Then it excretes all that stuff as urine.

The urinary system consists of the kidneys (the organs that filter the body's blood) and the system of tubes and structures that transport, store, and conduct urine from the body. These are the ureters, the bladder, and the urethra. This chapter describes each of these structures and the role it plays in the function of this vital body system.

You're In: The Kidneys

The kidneys are paired, bean-shaped organs located against the back wall of the upper abdominal cavity. Each kidney is about four to five inches long. The upper surface (or *superior pole*) of each kidney is at the level of the lower two ribs. The left kidney is slightly higher than the right kidney. Sitting on top of each superior pole are the *suprarenal* or *adrenal glands* (refer to Chapter 18 for more on how these work).

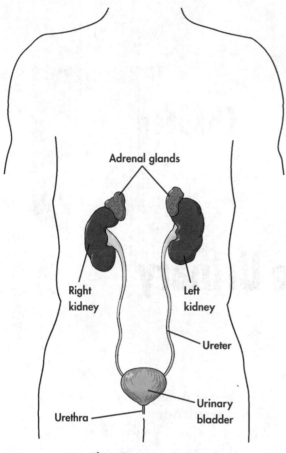

Adrenal glands

Right kidney

Left kidney

Ureter

Urinary bladder

Urethra

The urinary system.

The *hilum* or concave regions of the kidneys face each other. Within the hilum is a fat-filled space called the *renal sinus*. Here blood vessels and nerves enter and leave the kidney. The kidney is surrounded by a fibrous covering membrane called the *renal capsule*.

Body Language

Renal is a Latin word that refers to anything of or near the kidneys.

Inside, the kidney is organized into an outer *cortex*, beneath the renal capsule, and an inner *medulla* region. The medulla is made up of several cone-shaped structures called *renal pyramids*. There are about 7–15 pyramids per kidney. The bases of the pyramids are directed toward the cortex. Their tips (apices) are directed toward the renal sinus. The apex of a renal pyramid is called a *renal papilla*. Portions of the cortex, called *renal columns*, extend inward between the renal pyramids. Urine is formed along tubules located in both the cortex and the medulla (discussed later in this chapter).

As urine is formed, it leaves the renal pyramids at the renal papillae. Then it enters "cup-like" structures, the *minor calyces* (calyx, singular), that enclose the renal papillae. The minor calyces form the beginning of the urinary tube system located within the renal sinus. Several minor calyces join to form two to three major calyces. The major calyces then join to form the funnel-shape *renal pelvis*. The renal pelvis narrows at the hilum and continues downward along the posterior abdominal wall toward the pelvis as the *ureter*. The ureter is a narrow, muscular tube that transports urine by a series of wave-like contractions along the tube from the kidney to the urinary bladder. The kidney is enclosed by two different layers of fat to protect and support it during daily movements and activities.

The kidneys receive about 20 percent of the blood that is pumped out of the heart each minute. This blood arrives within the renal arteries. That means that the entire volume of circulating blood in the body is filtered through the kidneys every five minutes! Newly filtered and "cleansed" blood leaves the kidney within the renal veins, which are tributary to the inferior vena cava.

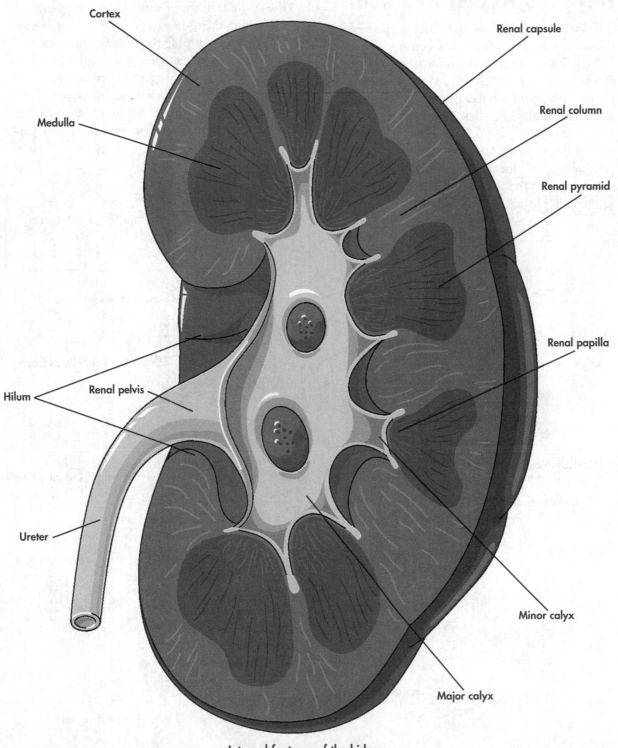

Internal features of the kidney.

Have You Met My Nephron?

The functional unit of the kidney is the *nephron*. There are over one million nephrons in each kidney. Each nephron is a microscopic blood-filtering unit. It consists of a vascular part, the *glomerulus*, and a tubular part of narrow, highly folded and coiled tubules. Together, these parts selectively filter plasma fluid; absorb glucose, amino acids, water, and salts; and secrete waste products of cell metabolism and other unwanted substances, in the form of urine. Most of the kidney's nephrons are in the cortex.

A *renal corpuscle* consists of a tuft of capillaries called the *glomerulus*. This is intimately enclosed in a cup-like structure called *Bowman's capsule*. After entering the kidney, each renal artery divides into smaller and smaller branches. Tiny arteries in the cortex give rise to *afferent arterioles*, from which arise the *glomeruli*. These capillaries are unique because they have openings that allow plasma fluid to readily pass through them.

Body Language

Nephro- is a Greek word that means "kidney." Both terms are used a lot when describing parts of the kidney.

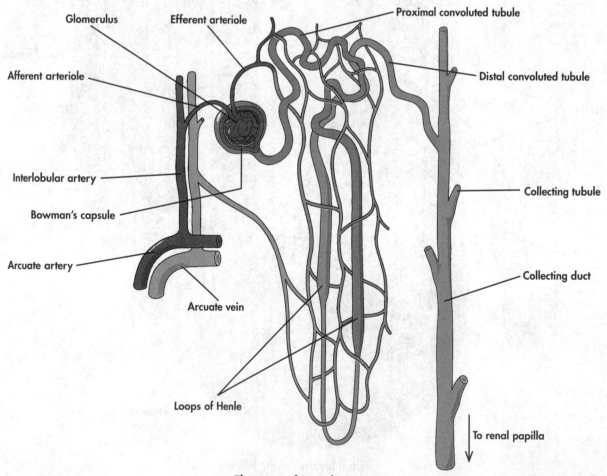

Glomerulus

Efferent arteriole

Proximal convoluted tubule

Afferent arteriole

Distal convoluted tubule

Interlobular artery

Bowman's capsule

Collecting tubule

Arcuate artery

Collecting duct

Arcuate vein

Loops of Henle

To renal papilla

The parts of a nephron.

The portion of Bowman's capsule that attaches to the glomerulus consists of specialized cells called *podocytes* that restrict the size of molecules and substances that can escape the capillaries and enter this capsule. The filtered fluid is referred to as the *glomerular filtrate*. It is similar in composition to the blood plasma but without its proteins because they cannot cross between these two spaces.

Bodily Malfunctions

Normally, the fluid filtered by the glomeruli lacks proteins because these molecules are too large to pass through the filtration barrier produced by the capillaries and Bowman's capsule. In certain diseases, such as diabetes, this barrier becomes less able to restrict larger molecules, so they are released into the urine. The presence of abnormal amounts of protein in the urine is called proteinuria.

Foot Notes

The kidneys produce about 1.5 liters of urine each day. It's normally a clear, yellowish fluid that consists mostly of water (about 95 percent) and urea, salts, and minerals. It is also sterile, meaning that it is free of bacteria and other microorganisms. The presence of the latter indicates a urinary infection.

Ureters: Shortcut from the Kidneys to the Bladder

The ureters are the tubes that connect the kidneys to the bladder. They leave the kidneys at the hilum as continuations of the renal pelvis. They are about 5 millimeters in diameter and 30 centimeters long. They pass downward along the muscles of the posterior abdominal wall,

cross the common iliac arteries, and then enter the pelvis.

The ureters enter the back side of the bladder and pass through the wall diagonally. When the bladder fills or contracts during urination, the opening of the ureter closes, preventing backflow of urine up the ureter.

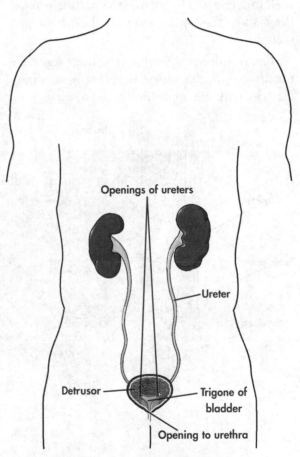

Openings of ureters

Ureter

Detrusor

Trigone of bladder

Opening to urethra

The ureters and urinary bladder.

Bodily Malfunctions

Kidney stones are hardened clumps of minerals that form (usually spontaneously) in the kidney. Occasionally, kidney stones that leave the kidney may get stuck within the ureter or at the point where the ureter enters the bladder wall. Dilation or extension of the ureter due to a stone is extremely painful and requires medical care.

Waiting to Go: The Urinary Bladder

The urinary bladder is a hollow muscular organ in the pelvis. Its smooth muscle wall is lined by mucosa cells with a covering of transitional epithelium. These cells change their shape when the bladder fills up, to accommodate the larger volumes.

The bladder is a temporary storage reservoir for urine until the person decides (or, on a long car trip, gets the opportunity) to empty it.

Ahh, Sweet Relief: The Urethra

The urethra is the passageway that conducts urine from the bladder out of the body. The urethra is short in the female—about 1½ to 2 inches long. It opens into the vestibule of the vagina. Only urine passes through the urethra in females.

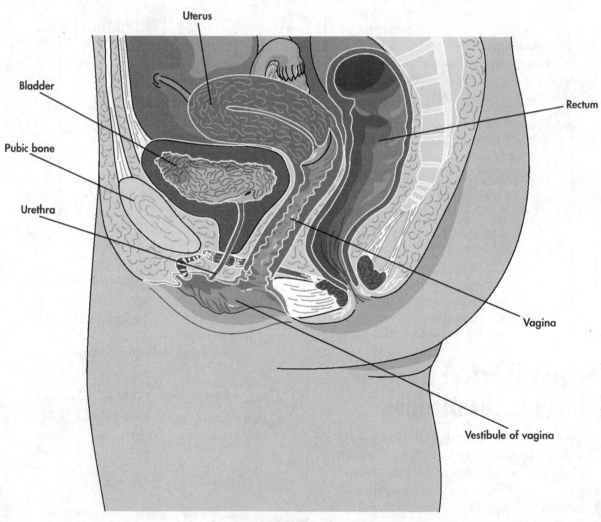

Uterus

Bladder

Pubic bone

Urethra

Rectum

Vagina

Vestibule of vagina

The female urethra.

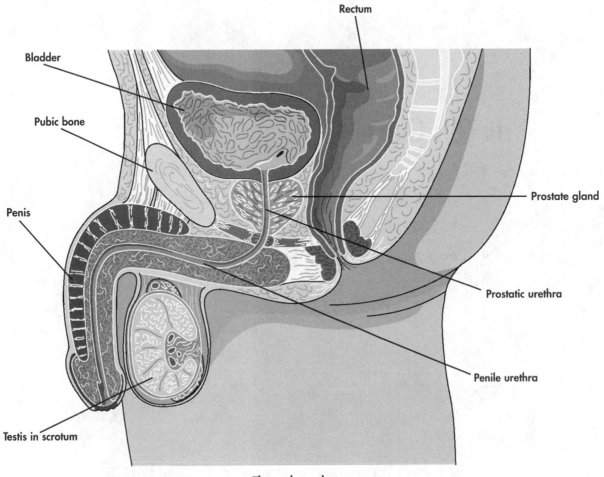

Rectum

Bladder

Pubic bone

Penis

Prostate gland

Prostatic urethra

Penile urethra

Testis in scrotum

The male urethra.

The urethra is considerably longer in the male—about 6 to 7 inches. From the bladder, it passes through the prostate gland before entering the penis. In addition to urine, semen also passes through the urethra during ejaculation (but never both at the same time).

The Least You Need to Know

◆ The urinary system rids the blood of metabolic waste products and excess water as urine.

◆ The functional unit of the kidney is the nephron, which filters blood plasma, absorbs water and salts, and secretes unwanted substances for disposal in the urine.

◆ Ureters transport urine from the kidney to the bladder.

◆ The urethrae are passageways that conduct urine from the bladder to the outside of the body. They are different in males and females.

In This Chapter

- ◆ The parts of the male reproductive system

- ◆ The production of sperm cells and the pathway of ducts they follow from the testes to the urethra

- ◆ The accessory glands and the substances they produce

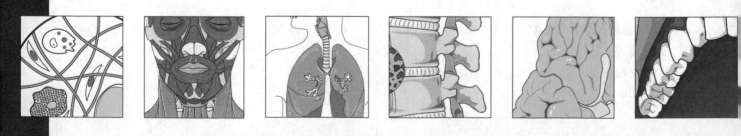

Chapter 12

The Male Reproductive System

The term "reproduction" usually means to produce an identical copy of an original, as in making a copy of a check or important document. While we may be "one of a kind" as individuals, we are never identical to the parents who created us. Reproduction in the biological sense refers to the creation of offspring of the same species. It is understood that such offspring will share some or many of the characteristics of their parents because each contributes equal amounts of genetic material through their reproductive cells. So, we are then likenesses of both our parents rather than an exact copy of either of them.

The next two chapters deal with the reproductive systems of both sexes. The centerpiece of each are the gonads, or sex glands, that produce the reproductive cells. Also important are the associated ducts, accessory glands, and supportive organs that enable humans to procreate. In this chapter, you will learn about the male reproductive system. You learn about the female reproductive system in Chapter 13.

Testes: One, Two...

The structures of the male reproductive system are located both outside and inside the body. The testes (or testicles) are soft, oval structures inside a pouch-like sac of skin called the *scrotum*. The scrotum hangs downward beneath the base of the penis.

The testes develop inside the abdominal cavity in the male fetus. Gradually they descend along the posterior wall of the abdomen toward the groin. During the last two months before birth, they pass through an opening in the wall of the abdomen, called the *inguinal canal*, and enter the scrotum. The placement of the testes in the scrotal sac outside the body is important for future sperm cell production and fertility. The testes must be about 4° Fahrenheit cooler than the body's temperature in order to produce sperm.

Bodily Malfunctions

Cryptorchidism or "hidden testicle" occurs in about 3 percent of full-term male newborns. It's the most common genital problem observed at birth. In this condition, a testis (or testes) has not completed its journey into the scrotum. Often, it will descend over the next few months, but a doctor must monitor its progress. Failure of the testis to enter the scrotum may lead to reduced fertility and an increased risk for testicular cancer later in life.

The scrotum is divided into two compartments by a partition of connective tissue that separates each testis. The *scrotal wall* is unique in that it contains smooth muscle that changes the size and shape of the scrotum to regulate the temperature of the testes. Contraction of the muscle when the scrotum is cold gives the skin a wrinkled appearance and increases its thickness to reduce heat loss. In contrast, when external temperatures are high, the muscle relaxes, the scrotum enlarges, and the skin becomes smooth and thin to promote heat loss. This activity is assisted by a muscle that brings the testes closer to or further from the body.

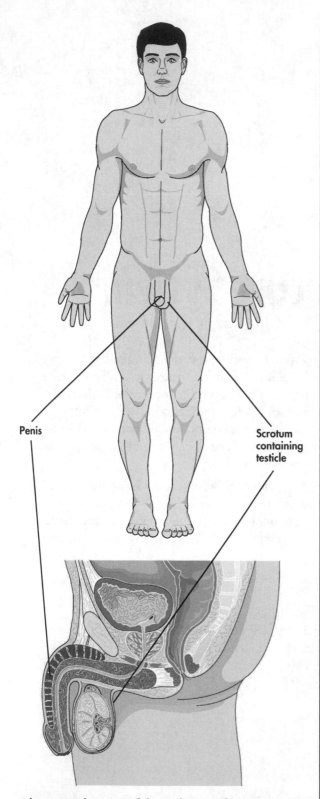

Penis

Scrotum containing testicle

The external organs of the male reproductive system.

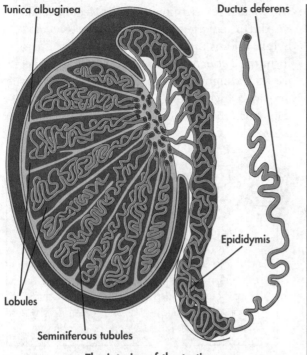

Tunica albuginea

Ductus deferens

Epididymis

Lobules

Seminiferous tubules

The interior of the testis.

The testes of an adult male are about two inches long and one inch around. Each testis is enclosed by a white fibrous capsule called the *tunica albuginea* (think of it as a tunic for the testicles). Fibers from this capsule project into the testis and subdivide it into approximately 250 lobules. Each lobule contains one to four highly coiled tubules called *seminiferous tubules*. This is where sperm cells are produced.

Body Language _____

A **tubule** is a thin, elongated channel or tube in the body used to conduct cells or fluids.

Each seminiferous tubule is about 30 inches long. These tubules join a few short tubules that leave the testis and enter the epididymis.

The testis has two basic functions: to produce sperm cells and to produce the hormone testosterone. Testosterone is necessary for the formation of sperm cells and the development of secondary sex characteristics in the body, such as body and facial hair, a deeper voice, increased muscle mass, and body strength.

Foot Notes _____

The process by which sperm cells are produced is called *spermatogenesis*. The process is controlled by the hormone testosterone. Spermatogenesis occurs within the seminiferous tubules. It begins at puberty and continues throughout the life of the individual, although sperm cell production decreases as a man ages. On average, 70–100 million sperm cells are produced each day.

Sex Hormone Production

The principal sex *hormone* produced by the testis is testosterone. This hormone is necessary for the production of sperm cells and for development of the secondary male sex characteristics. Production of this hormone begins at puberty.

Body Language _____

Hormones are chemicals produced by endocrine cells. They are carried in the bloodstream and regulate the activities of other distant cells.

The Epididymis and Ductus Deferens

Once spermatozoa are formed, they leave the testes and enter a highly coiled tube called the *epididymis*. The epididymis is located along the back surface of the testis. It is about 15 feet long and consists of a head, body, and tail portion.

Newly formed sperm are not able to fertilize the female egg and must first undergo a maturation process during their transit in the epididymis. They are temporarily stored in the tail of the epididymis.

During sexual arousal, the spermatozoa enter and are transported along the *ductus* or *vas deferens*. The ductus has a narrow channel and a thick muscular wall. This wall strongly propels the cells along the duct and toward the urethra. The ductus deferens is a component of the *spermatic cord*, which also includes the *testicular artery*, the *pampiniform plexus* of veins, the *cremasteric muscle*, nerves, and lymphatic channels. The spermatic cord connects the testes to the lower abdominal wall.

The ductus deferens passes through the abdominal wall and into the abdominopelvic cavity. It courses downward behind the bladder and medial to a seminal vesicle. Near its end, the ductus enlarges and is called the *ampulla* before it joins the duct of the *seminal vesicle* to form the *ejaculatory duct*. The ejaculatory duct passes through the prostate gland to open in the urethra, which is surrounded by the prostate. This portion of the urethra is called the *prostatic urethra*.

Foot Notes

A vasectomy is a permanent birth-control method for men. It is a simple surgical procedure where the ductus deferens is accessed through a small opening in the scrotum. It is cut and tied so that sperm can no longer pass into the penis. This procedure does not affect the production of male hormones by the testes or significantly change the volume of semen.

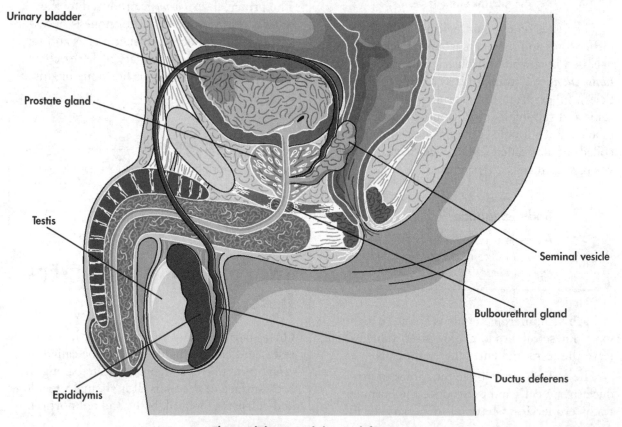

The epididymis and ductus deferens.

Accessorize Your Genitals: The Accessory Glands

Sperm become highly active when they mix with the fluids produced by the accessory glands (such as the *seminal vesicles*, the *prostate gland*, and the *bulbourethral glands*) to form the ejaculate. This fluid is called seminal fluid or semen. It contains abundant sources of energy and activator substances, and has a slightly basic pH to counter any residual urine within the male tract and the acidity of the female vagina.

Seminal Vesicles

The seminal vesicles are two narrow, sac-like structures about five to six inches long located behind the bladder. They do not store sperm. Rather, they produce about 70 percent of the volume of semen. This fluid is rich in fructose, an energy source, and other sperm activators. The duct of a seminal vesicle joins the ductus deferens to form an ejaculatory duct that passes through the prostate gland and empties into the prostatic urethra.

The Prostate Gland

The prostate gland is shaped like a walnut. It surrounds the urethra as it leaves the bladder. The prostate is located in front of the seminal vesicles and the ends of the ductus deferens. It's in front of the rectum as well. The prostate consists of firm, dense tissue within which are numerous glands. These glands produce the remaining volume of the seminal fluid. Their ducts open into the prostatic urethra and empty their contents during ejaculation. They produce a thin, milky, alkaline solution that enables the sperm to swim better (which makes them more likely to fertilize the female's egg).

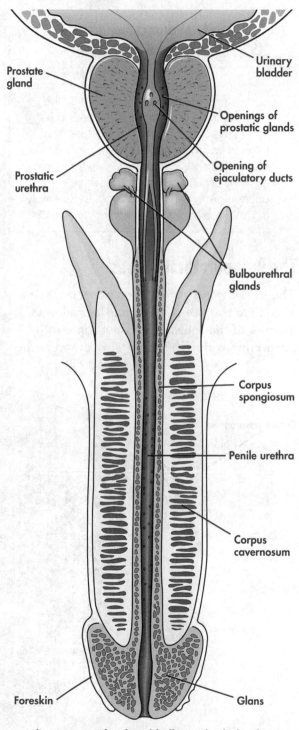

The prostate gland and bulbourethral glands.

Bodily Malfunctions

Prostate cancer is the second most common type of cancer among men in the United States, next to skin cancer. Benign prostatic hypertrophy, on the other hand, is a common non-cancerous enlargement of the prostate in older men. Glands within the prostate overgrow and narrow or obstruct the urethra and reduce the flow of urine out of the bladder. A frequent urge to urinate and incomplete emptying of the bladder are common symptoms.

The Bulbourethral Glands

The *bulbourethral glands*, also known as *Cowper's glands*, are two pea-sized glands located near the root of the penis. Their ducts open into the urethra within the penis (also known as the penile urethra). During sexual arousal, these glands secrete a clear, slippery alkaline fluid that lubricates the urethra and neutralizes any residual acid urine in anticipation of ejaculation.

The Penis

The penis is the public spokesperson of the male reproductive system. It carries urine from the bladder out of the body. It is also used in sexual intercourse to transfer sperm to the vagina. To accomplish the latter, the penis must be erect.

The root of the penis attaches it to the body wall. The body or shaft of the penis hangs down from the body wall when it is flaccid. The body of the penis is cylindrical but expands at its end as the glans (or head) of the penis. The glans is covered by a fold of skin, called the *foreskin* or *prepuce*, which is removed in a circumcision.

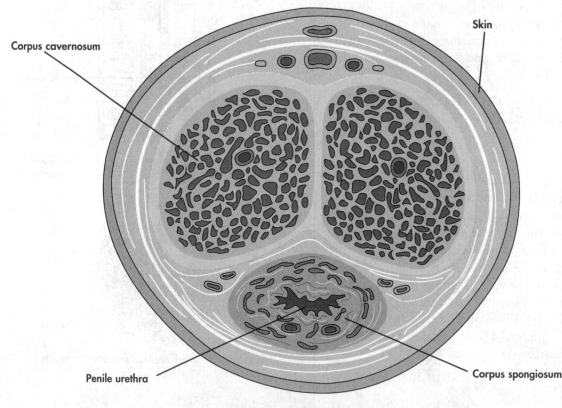

Corpus cavernosum

Skin

Penile urethra

Corpus spongiosum

A cross section of the penis.

The body of the penis consists of three cylindrical masses of sponge-like tissue enclosed by tough connective tissue and skin. There are two pairs of *corpora cavernosa* and a single *corpus spongiosum* that contains the urethra. The corpus spongiosum expands distally as the glans of the penis. The penile urethra is the passageway for urine and semen. During sexual arousal, arteries within the corpora cavernosa deliver blood to the tissue spaces of these erectile bodies, causing them to fill with blood. As a result, the penis becomes stiff and erect so that it is capable of penetrating the vagina during intercourse.

The Least You Need to Know

- The testes have a dual function: to produce sperm cells and to produce the sex hormone, testosterone.
- The ducts that transport the sperm from the testes to the urethra are the epididymis, the ductus deferens, and the ejaculatory duct.
- The seminal vesicles, prostate gland, and bulbourethral glands are accessory glands that produce fluids that mix with the sperm to increase their capacity to fertilize an egg.
- The penis has a dual function to carry urine out of the body and to deliver sperm into the female tract during intercourse.

In This Chapter

- ◆ The parts of the female reproductive system
- ◆ The production of egg cells and how they get to the uterus
- ◆ The mammary gland and features of the breast

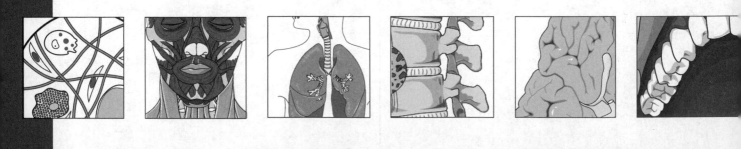

Chapter 13

The Female Reproductive System

Any way you look at it, mothers are special people. By the time an infant is born, his or her mother has already invested a lot of herself in this miracle of life. Apart from the contribution of the male in the form of sperm cells, her reproductive system cradles and nurtures the development of the fetus for many months until it is born. Her body continues to provide for the newborn's nutrition by the milk from her breasts, sometimes until the baby is a year old.

The function of the female reproductive system is to produce reproductive cells, the eggs, which are made available to sperm cells for fertilization, and to support the development of the fertilized egg into a fetus. In addition, this system produces sex hormones, estrogen and progesterone, that control egg cell production and the organs of the female reproductive system. It also plays a role in the development of secondary sex characteristics, such as breast development, body distribution of fat and hair, high-pitched voice, and the menstrual period. This chapter describes the parts of the female reproductive system and how each works to support the development of a sexually mature individual.

The female reproductive system consists of both external and internal organs. I will begin by describing the internal organs.

Internal Reproductive Organs

Most of the female reproductive magic happens internally. This section takes a look at the organs on the inside: the ovaries, the uterine tubes, the uterus, and the vagina. The next section goes into detail about the external organs.

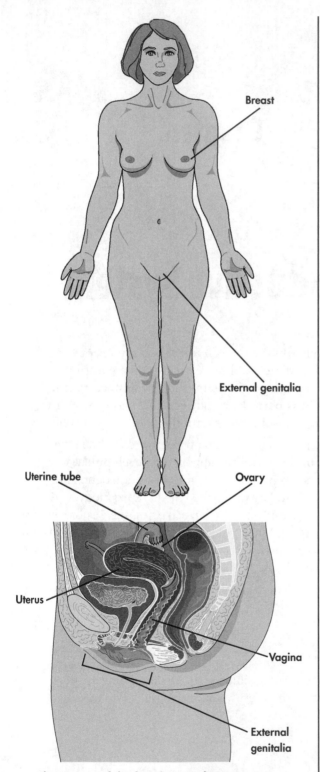

The organs of the female reproductive system.

Madame Ovary

The organs that produce the egg cells (*oocytes*) are the ovaries. The two ovaries are located in the pelvic cavity next to the uterus and beneath the Fallopian or uterine tubes.

Body Language _____

Oocytes, or ova, are the egg cells produced by the female ovaries.

In an adult female, these are almond-shaped structures that are about three centimeters long, two centimeters wide, and one centimeter thick. The surface of each ovary is lined by a special cuboidal tissue known as germinal epithelium. Beneath this is a fibrous connective tissue capsule called the *tunica albuginea* (remember from Chapter 12 that the male testicles have a similar "tunic"). The adult ovary consists of an outer cortical region containing *follicles* (predeveloped egg cells) at various stages of development and an inner medulla that is rich in blood vessels. Ovarian follicles consist of an oocyte surrounded by supportive cells called *follicular cells*.

Egg cell production actually begins within the ovary during fetal development. At birth, a female has about two million eggs in her ovaries—all the egg cells she will ever have. Many of these will die so that by puberty, only about 300,000 remain. But that is plenty for her reproductive needs. A woman ends up ovulating only approximately 400 eggs in her lifetime.

The follicles inside the ovaries are inactive until puberty. Then, under the influence of follicle stimulating hormone (FSH) from the pituitary gland, clusters of 15–20 follicles begin to develop on one of the ovaries. This process normally repeats itself about every 28 days (usually alternating from one ovary to the other each month) for the reproductive lifetime of the woman. Of these clusters, only one egg-bearing follicle ripens and advances to be released from the ovary. This process is called *ovulation*.

Foot Notes

Oogenesis is the production of female sex cells, the egg or oocyte. Unlike in the male, this production is not continuous, but occurs on a monthly basis. And when the female runs out of eggs (generally 30 or 40 years after puberty), ovulation stops and menopause begins.

Catch and Release: The Uterine Tubes

The *uterine tubes*, also called *fallopian tubes* or *oviducts* ("egg tubes"), are narrow, muscular tube structures that extend from the uterus to each ovary. Each tube is about five inches long. Near the ovary, the tube flares into a funnel-shaped end called the *infundibulum*. This part is open to the peritoneal cavity and is surrounded by numerous delicate, finger-like extensions called *fimbriae*.

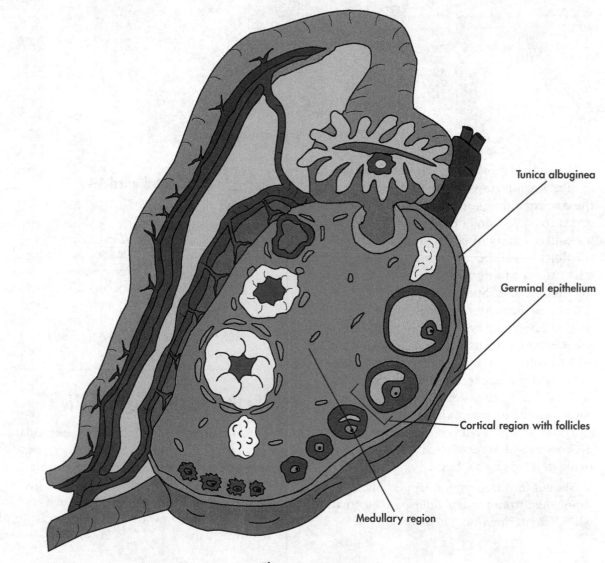

The ovary.

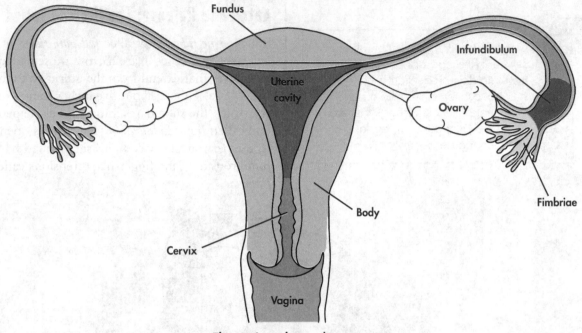

The uterine tubes and uterus.

When ovulation occurs, the fimbriae reach up like a hand and sweep the released egg into the corresponding tube. Once it enters, the egg is conveyed along the lining of the tube by wave-like beating of cilia and muscular contraction toward the uterus. Fertilization, when a sperm unites with the egg, occurs within the uterine tube. Sperm cells introduced into the vagina during intercourse "swim" into the uterine cavity and eventually enter the uterine tubes, where they finally meet up with the egg. After fertilization, it takes three to four days for a fertilized egg to reach the uterine cavity and another three or four days before it burrows into the uterine wall. By this time, rapid cell division has produced a structure called the blastocyst, which becomes an embryo and eventually develops into a fetus.

Should fertilization not happen, the egg is simply lost from the body during menstruation when the uterine lining is shed.

Bodily Malfunctions

Occasionally, fertilized eggs implant outside the uterus. Such pregnancies are referred to as ectopic pregnancies. The most common ectopic pregnancy is a tubal pregnancy, meaning that implantation has occurred in the lining of the uterine tube. If a tubal pregnancy is not detected and surgically removed soon enough, the tube can rupture. Such a rupture can be life-threatening.

The Uterus: The Cradle of Life

The uterus, or womb, is a hollow, muscular organ that looks like an upside-down pear. It is about three inches long, two inches wide, and an inch thick. The uterus is located in the center of the pelvic cavity behind the urinary bladder and in front of the rectum. It consists of the following parts:

◆ **Fundus.**   The domed part of the uterus above the entrance of the uterine tubes

◆ **Body.**   The major part of the uterus

◆ **Cervix.**   The lowermost portion of the uterus that projects into the vagina

The body of the uterus consists of two principal layers:

◆ **Endometrium.**   This is the internal lining of the uterine cavity. It contains abundant glands and blood vessels. This layer thickens each month under the hormonal influence of estrogen and progesterone (produced by cells in the ovary) to prepare for the potential implantation of a fertilized egg (now a blastocyst) into its wall. If *implantation* does occur, this layer forms part of the placenta that nourishes and supports continued growth of the embryo and fetus. If fertilization does not occur, a portion of this lining (next to the uterine cavity) breaks down and is shed through the vagina as the menstrual flow each month. Following menstruation, this lining is regrown from the portion of the endometrium attached to the muscle wall.

◆ **Myometrium.**   This is the thick muscular wall of the uterus. It grows nearly 20 times its original size during pregnancy. Its strong contractions force the fetus out through the cervical canal toward the vagina during birth.

Body Language

Implantation is the term that describes the process by which the blastocyst (the early conceptus produced by rapid cell division of the fertilized egg) burrows into the lining of the uterus. Cells from both the blastocyst and uterus help form the placenta that provides the developing embryo and fetus with nutrients and blood.

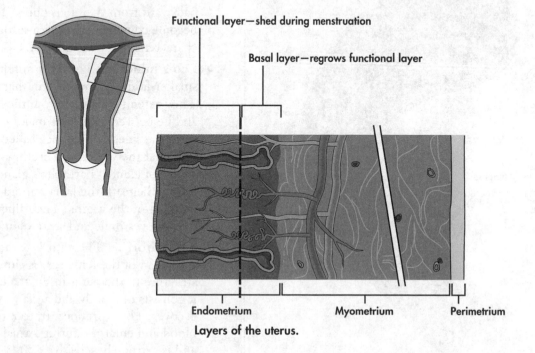

Functional layer—shed during menstruation

Basal layer—regrows functional layer

Endometrium   Myometrium   Perimetrium

Layers of the uterus.

The Vagina

The *vagina* is a hollow, distensible fibromuscular structure about three to five inches in length. It opens at one end with the vestibule, and surrounds and attaches at its opposite end to the cervix of the uterus.

Body Language

Vagina means "sheath," and it gets its name from the sheath-like relationship it has with the penis during sexual intercourse.

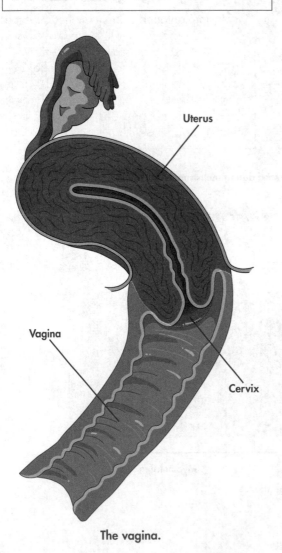

Uterus

Vagina

Cervix

The vagina.

The vagina functions as the passageway for menstrual blood from the uterus out of the body, as the lower end of the birth canal during delivery of a fetus, and to receive the penis during intercourse.

External Genitalia

The external genitalia, also called the vulva, are the external organs of the female reproductive system. They guard the openings of the urethra and vagina at the vestibule and prevent bacteria from getting into the bladder and uterus. The external genitalia consist of the following structures:

- ◆ **Mons pubis.** The mons pubis is a rounded prominence of fatty tissue overlying the symphysis pubis, where the two pubic bones meet. It becomes covered with hair after puberty.

- ◆ **Labia majora.** Labia means "lips." These are two fat-filled folds of skin that extend backward from the mons pubis. They, too, become covered by hair as a secondary sex characteristic during puberty.

- ◆ **Labia minora.** The labia minora are two smaller folds of smooth skin that lack hair. They extend backward from the clitoris and lie between the labia majora. They form the sides of the space called the vestibule that the vagina and urethra open into. A pair of glands, Bartholin's glands, are located beneath the labia minora near the opening of the vagina. Their ducts open into the vestibule and secrete mucous.

- ◆ **The clitoris.** The clitoris is the erectile organ of the female. It is similar in structure to the penis in the male, in that it consists of a body and a glans, which is covered by a prepuce. It, too, fills with blood and enlarges during sexual arousal and is extremely sensitive.

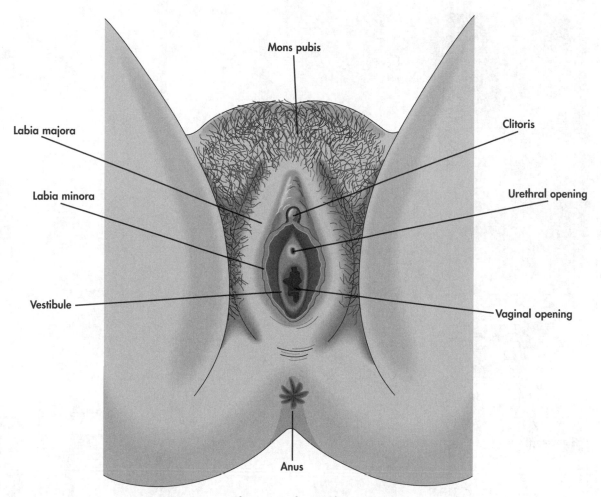

Mons pubis

Labia majora

Labia minora

Vestibule

Clitoris

Urethral opening

Vaginal opening

Anus

The external genitalia.

Thanks for the Mammaries!

The mammary glands are modified sweat glands that produce milk for the newborn infant. These glands are located within the breasts that attach to the front of the upper chest wall. The breasts also contain abundant fat and connective tissue that support this glandular tissue and give shape to the breasts. Most of the nonpregnant breast consists of fat.

There are 15–20 lobes of glandular tissue in each breast. Leaving each lobe is a large

lactiferous duct. All of these ducts converge on the nipple. The lactiferous ducts carry the milk produced by cells within each lobe. Beneath the nipple, each duct enlarges as a lactiferous sinus, which stores milk during lactation. The ducts narrow again and open onto the nipple.

The nipple is a raised, conical structure that contains smooth muscle and a rich sensory nerve supply. This makes it sensitive to touch, causing it to become erect, as occurs during suckling by the newborn. Surrounding the nipple is a circular pigmented area called the areola.

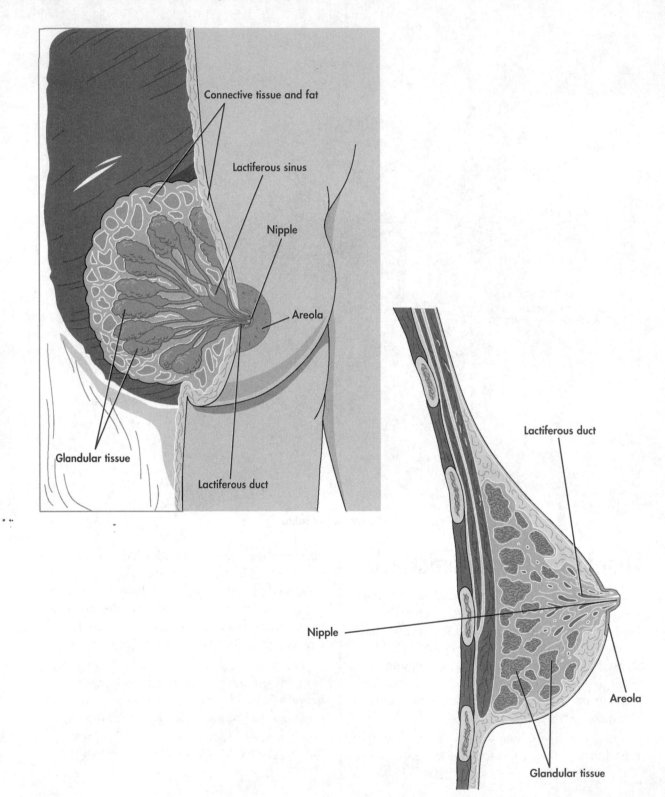

The breast.

Anterior View of the Skull

(Image Courtesy of Indiana University School of Medicine Office of Visual Media.)

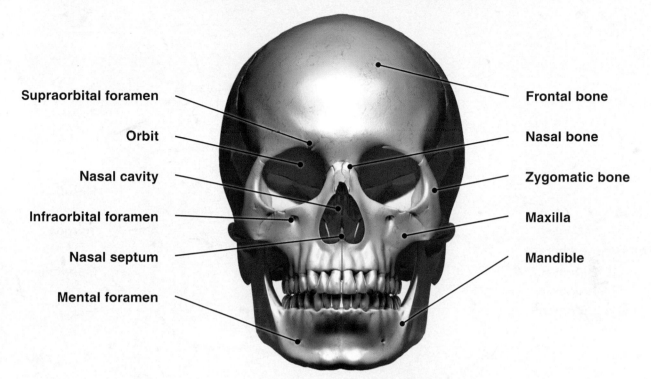

Supraorbital foramen

Orbit

Nasal cavity

Infraorbital foramen

Nasal septum

Mental foramen

Frontal bone

Nasal bone

Zygomatic bone

Maxilla

Mandible

Anterolateral View of the Skull

(Image Courtesy of Indiana University School of Medicine Office of Visual Media)

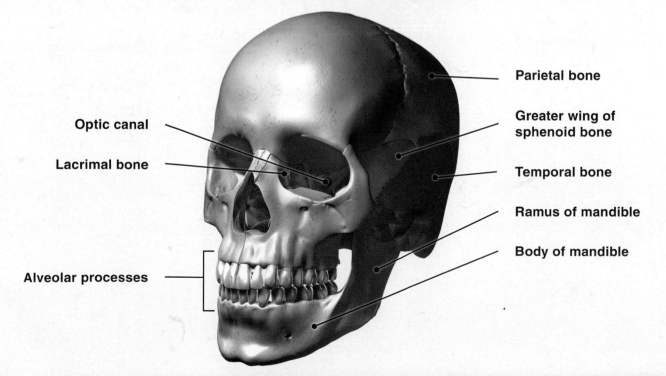

Optic canal

Lacrimal bone

Alveolar processes

Parietal bone

Greater wing of sphenoid bone

Temporal bone

Ramus of mandible

Body of mandible

Sagittal View of the Neck and Pharynx

(Image Courtesy of Indiana University School of Medicine Office of Visual Media.)

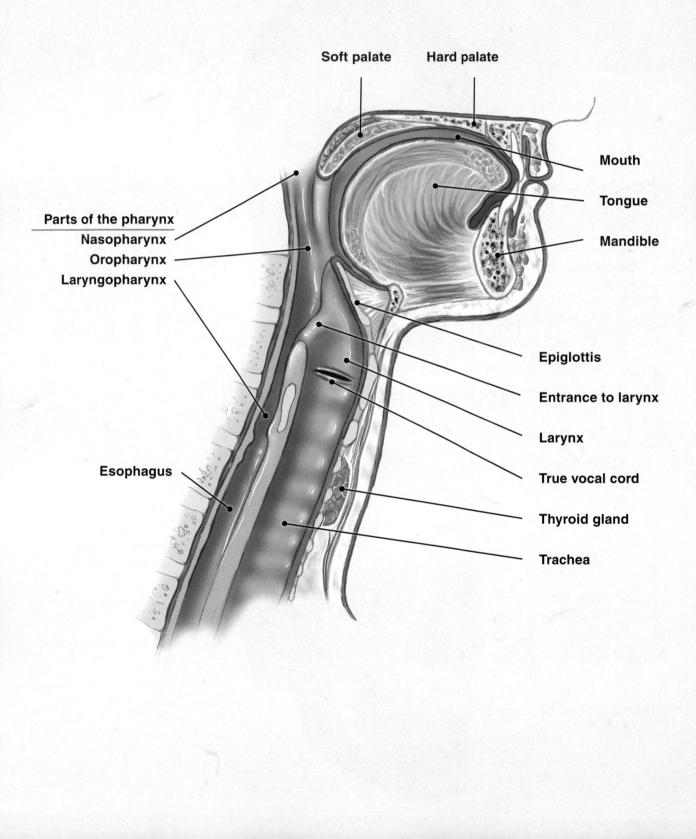

Soft palate

Hard palate

Mouth

Tongue

Parts of the pharynx

Nasopharynx

Oropharynx

Laryngopharynx

Mandible

Epiglottis

Entrance to larynx

Larynx

True vocal cord

Esophagus

Thyroid gland

Trachea

The Heart and Great Vessels

(Top Images © 2007 Wolters Kluwer Health \ Lippincott Williams and Wilkins.)
(Bottom Image Courtesy of Indiana University School of Medicine Office of Visual Media.)

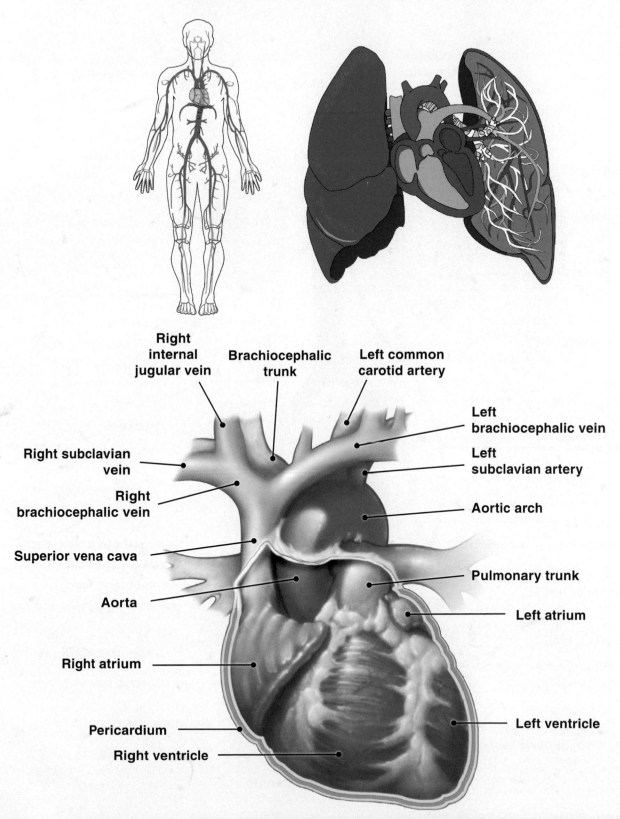

Right internal jugular vein

Brachiocephalic trunk

Left common carotid artery

Left brachiocephalic vein

Right subclavian vein

Left subclavian artery

Right brachiocephalic vein

Aortic arch

Superior vena cava

Pulmonary trunk

Aorta

Left atrium

Right atrium

Pericardium

Left ventricle

Right ventricle

The Male Reproductive System

(Image Courtesy of Indiana University School of Medicine Office of Visual Media.)

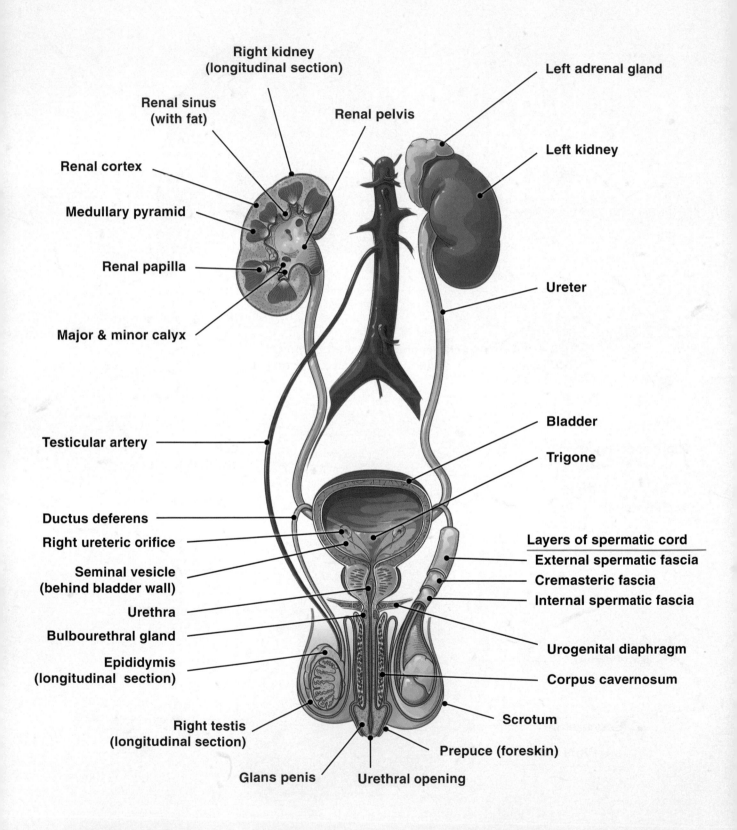

Right kidney (longitudinal section)

Left adrenal gland

Renal sinus (with fat)

Renal pelvis

Left kidney

Renal cortex

Medullary pyramid

Renal papilla

Ureter

Major & minor calyx

Bladder

Testicular artery

Trigone

Ductus deferens

Right ureteric orifice

Layers of spermatic cord

External spermatic fascia

Cremasteric fascia

Seminal vesicle (behind bladder wall)

Internal spermatic fascia

Urethra

Bulbourethral gland

Urogenital diaphragm

Epididymis (longitudinal section)

Corpus cavernosum

Right testis (longitudinal section)

Scrotum

Prepuce (foreskin)

Glans penis

Urethral opening

The Female Reproductive System

(Image Courtesy of Indiana University School of Medicine Office of Visual Media.)

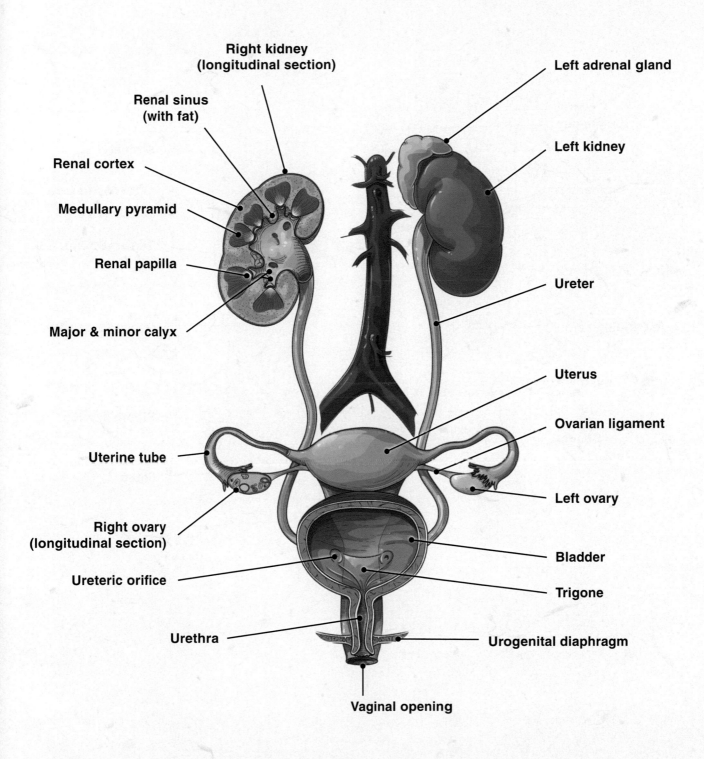

Right kidney
(longitudinal section)

Renal sinus
(with fat)

Renal cortex

Medullary pyramid

Renal papilla

Major & minor calyx

Uterine tube

Right ovary
(longitudinal section)

Ureteric orifice

Urethra

Left adrenal gland

Left kidney

Ureter

Uterus

Ovarian ligament

Left ovary

Bladder

Trigone

Urogenital diaphragm

Vaginal opening

The Gastrointestinal Tract

(Image Courtesy of Indiana University School of Medicine Office of Visual Media.)

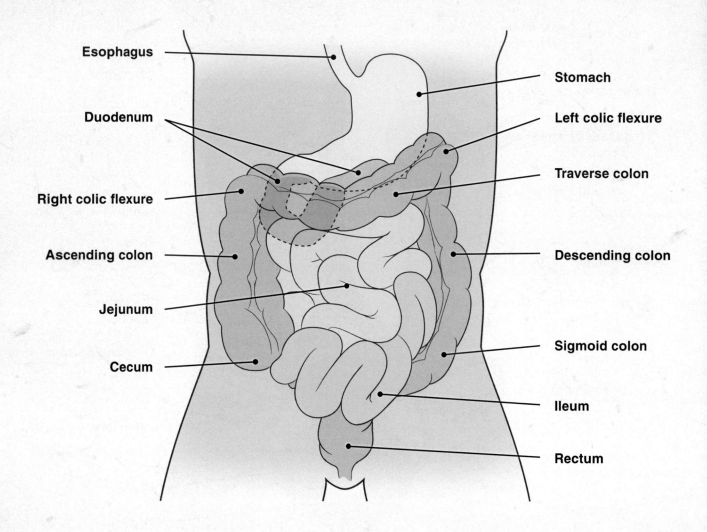

Esophagus

Duodenum

Right colic flexure

Ascending colon

Jejunum

Cecum

Stomach

Left colic flexure

Traverse colon

Descending colon

Sigmoid colon

Ileum

Rectum

The Spinal Cord and Meninges

(Left Image Courtesy of Indiana University School of Medicine Office of Visual Media.)
(Right Image © 2007 Wolters Kluwer Health \ Lippincott Williams and Wilkins.)

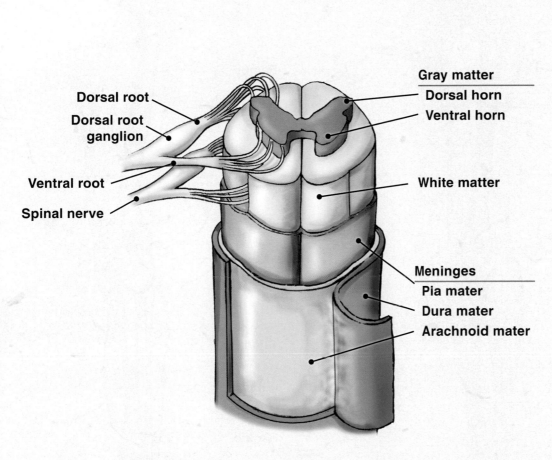

Dorsal root

Dorsal root ganglion

Ventral root

Spinal nerve

Gray matter
Dorsal horn
Ventral horn

White matter

Meninges
Pia mater
Dura mater
Arachnoid mater

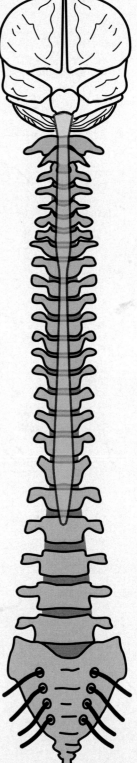

The Bones of the Upper Limb

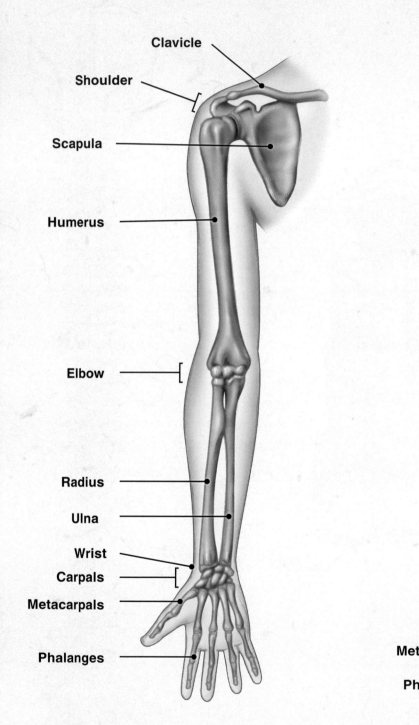

Clavicle

Shoulder

Scapula

Humerus

Elbow

Radius

Ulna

Wrist

Carpals

Metacarpals

Phalanges

The Bones of the Lower Limb

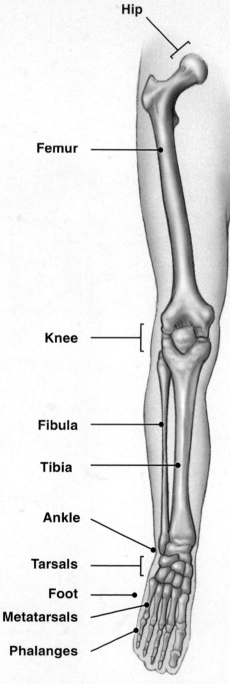

Hip

Femur

Knee

Fibula

Tibia

Ankle

Tarsals

Foot

Metatarsals

Phalanges

The tissue components and appearance of the breast are the same in both boys and girls before puberty. However, changes begin to take place in the female breast during the onset of puberty as part of the secondary changes to the body. Under the influence of estrogen, produced by cells within developing ovarian follicles, there is an increased accumulation of fat within the breasts and the glandular tissue begins to grow. Progesterone causes the duct system to develop in anticipation of pregnancy. The milk-producing apparatus, however, does not fully develop and become functional until the third trimester of pregnancy.

The Least You Need to Know

◆ The ovaries contain a limited number of follicles that contain egg cells (called oocytes). One oocyte is usually released each month under the influence of hormones.

◆ Oocytes are captured by fimbriae of the uterine tubes and carried within these tubes to the uterus. Along this course, they may become fertilized by sperm deposited by the male in the female tract.

◆ The uterus lining supports the implantation and development of the fertilized egg into an embryo and fetus. Its muscular wall pushes the fetus out through the vagina during birth.

◆ The mammary glands are components of the breasts. They are modified sweat glands that, under hormonal stimulation, develop as secondary sex characteristics and potential milk-producing organs for the newborn infant.

Part

3

Putting It All Together: Regional Context

Now that you know what the systems are and what they include, this part puts everything together and looks at it in a regional context. For example, the chapter about the back shows how all the back's various parts fit and function together: the spine and its joints, ligaments, and nerves.

But first, you'll learn about the various types of body movement and other terminology that will help you make sense of everything else you're about to learn.

In This Chapter

◆ Square one: the anatomical reference position

◆ The anatomical planes of the body

◆ The common anatomical terms used to describe the relationship, position, and movement of body structures

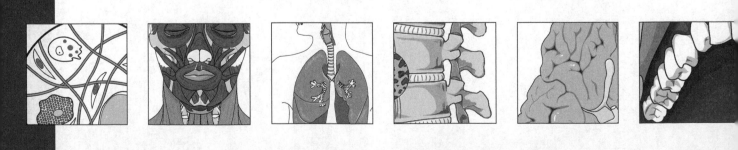

Chapter 14

The Language of Anatomy: Planes, Position, and Movement

Unless there are reference points and standardization in technical language, it becomes easy for otherwise intelligent people to get confused or unable to mentally follow what is being described. Imagine doctors and nurses and other health care professionals trying to describe the location and relationship of structures, diseases, or injuries in the body without a common reference language to use. There would be total chaos that could potentially injure patients. Fortunately, anatomy and medicine have had a very long history together. A universal descriptive language has evolved, which anatomists and clinicians worldwide learn during their education and training.

This chapter introduces the most common and essential elements of this anatomico-medical language. Knowing these terms enables you to better understand the study of anatomy.

Not Just Any Position: The Anatomical Position

All anatomic descriptions in the body are made in reference to something called the *anatomical position*, which is universally accepted and adopted by anatomists and clinicians alike. In this position, the person is standing erect, looking straight ahead, with arms at the sides and palms facing forward. The legs are together, with toes directed forward. Structures of the body are described in relationship to this position.

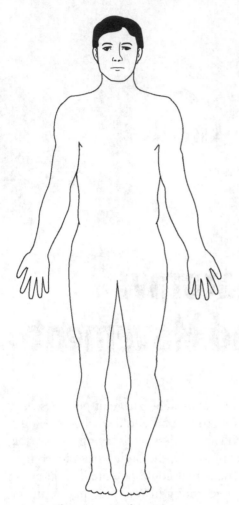

The anatomical position.

The anatomical position is slightly unnatural and is not the usual functional position of the upper limbs. Rarely would a person have their limbs facing this way. However, this is the position that allows a straightforward description of structures and relationships.

On a Different Plane

In anatomy, a *plane* is an imaginary line that divides the body into smaller sections for study. Doctors and technicians routinely view humans today in these various planes by using sophisticated radiologic imaging equipment, such as x-rays, computed tomography (CT), and magnetic resonance imaging (MRI) machines. These planes include the following:

◆ **Median or mid-sagittal plane.** This plane divides the body into equal left and right halves. Planes that are parallel to the median plane (also called the midline) are called *paramedian*, *sagittal*, and *parasagittal* planes.

◆ **Coronal or frontal plane.** This plane passes through the body at a right angle to the median plane. It divides the body into front (*anterior*) and back (*posterior*) parts.

◆ **Horizontal or transverse plane.** This plane passes through the body at right angles to the median and coronal planes and divides the body into upper (*superior*) and lower (*inferior*) parts.

Location, Location, Location: Anatomical Terms of Relationship or Position

A variety of terms, typically used as pairs of opposites, describe the location of a structure in relationship to another structure. Here are the main terms and some examples:

◆ **Anterior or ventral versus posterior or dorsal.** A structure that is anterior or ventral is closer to the front of the body than a structure that is posterior or dorsal, which lies closer to the back of the body. For example, the sternum (breastbone) lies anterior to the heart.

◆ **Superior or cranial versus inferior or caudal.** Superior means closer to the head or cranium. Inferior means closer to the soles of the feet. For example, the neck is superior to the chest, whereas the diaphragm is located inferior to the heart.

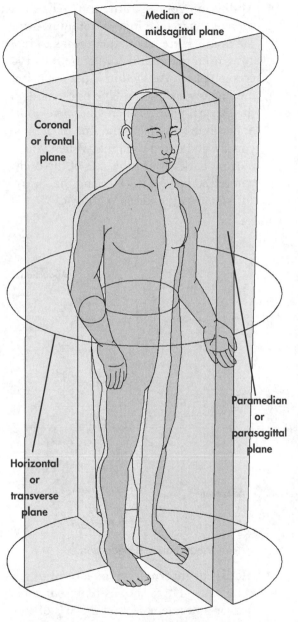

Anatomical planes.

- **Medial versus lateral.** Medial refers to something being closer to the midline of the body, whereas lateral refers to something further from the midline.

For example, the shoulder is lateral to the neck. The little finger is medial to the thumb. (This is tricky. Remember the anatomical position!)

- **Proximal versus distal.** Something that is proximal is closer to the body or some point of origin. Something that is distal is further from the body or some point of origin. For example, the shoulder is proximal to the elbow. We can easily say the same thing in a different way by flipping these terms around, like, "the elbow is distal to the shoulder." Saying "the elbow is proximal to the wrist," has the same meaning as "the wrist is distal to the elbow."

- **Superficial versus deep.** This one's a little more intuitive. Superficial means closer to the surface or skin of the body. Deep means further from the surface of the body. As in the preceding bullet, "the sternum (breastbone) lies superficial to the heart" means the same things as "the heart lies deep to the sternum."

Foot Notes

The beauty of using the anatomical position reference is that it helps avoid confusion. Regardless of whether you're standing on your head or lying on your back, reference to structures in the body is always made relative to the anatomical position. Even if you're doing a handstand, your head is still located superior or cranial to your neck.

Not so difficult when you think about it, right? Just remember that these pairs of terms are used as opposites and help pinpoint in your mind the location and relationship of structures.

And, Action!: Anatomical Terms of Movement

A variety of terms, again used as pairs of opposites, describe movements of the body at the joints. Here are the terms and some examples to illustrate them:

◆ **Flexion versus extension.** Flexion means bending a body part or decreasing the angle at a joint. Extension means straightening a body part or increasing the angle at a joint. For example, flexion of the spine is leaning forward to touch the toes. Extension of the spine is returning to the upright posture.

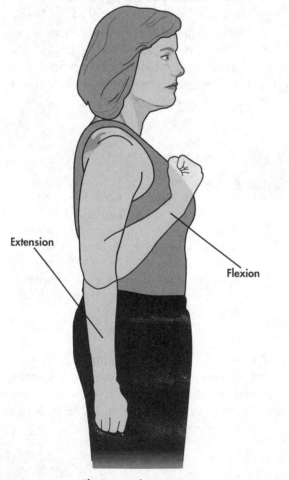

Flexion and extension.

◆ **Abduction versus adduction.** To abduct someone is to take him or her away from the home. Here, abduction means to move away from the midline of the body in the coronal plane. Adduction is the opposite: moving toward the midline of the body in the coronal plane. When you move your upper limb away from the side of the body, you are abducting the limb. When you move it back toward the side of the body, you are adducting it. (Hearing the "b" or "d" sound in the terms is critical.)

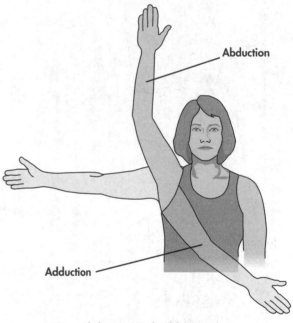

Abduction and adduction.

◆ **Medial or internal versus lateral or external.** These actions represent rotation or movement around the axis of a limb or long bone. Medial rotation moves the anterior surface of a limb toward the body; lateral rotation moves the anterior surface of a limb away from the body. You frequently use these movements without knowing it. When you first move your pen to the beginning of a sentence you want to write on a piece of paper, you have medially or internally rotated your arm. As you continue to complete this sentence, you have laterally or externally rotated your arm. Medial rotation of the lower limb causes the big toe to point inward (medially). The opposite action, lateral rotation, causes the big toe to point outward (laterally).

◆ **Circumduction.** Think of this as "around the world." Circumduction is the universal type of free movement at ball-and-socket joints, like the shoulder and hip. It's a circular movement that combines flexion, abduction, extension, and adduction.

◆ **Supination versus pronation.** These actions happen in the forearm. In the anatomical position, the forearm is in the supinated position and the bones of the forearm (the radius and ulna) are parallel to each other: the palms face forward (anteriorly). Pronation involves medial rotation of the radius across the ulna, like an "X", so that the palms face backward (posteriorly).

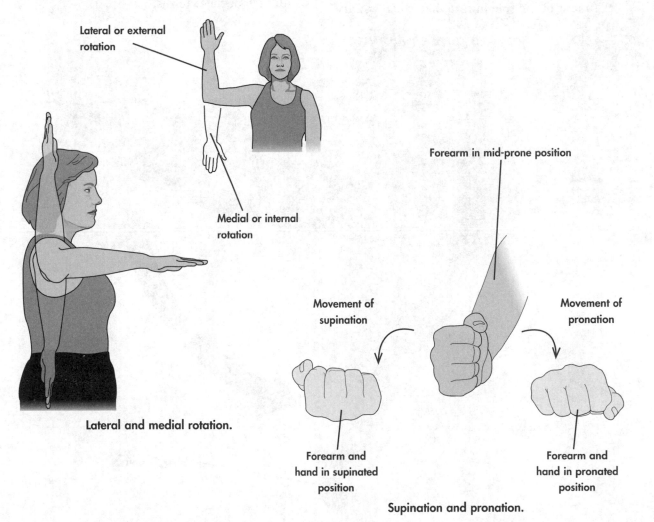

Lateral or external rotation

Medial or internal rotation

Lateral and medial rotation.

Forearm in mid-prone position

Movement of supination

Movement of pronation

Forearm and hand in supinated position

Forearm and hand in pronated position

Supination and pronation.

◆ **Dorsiflexion versus plantar flexion.** These describe movements at the ankle joint. Dorsiflexion is movement of the toes and foot toward the leg, like when a person stands on their heels. Plantar flexion is movement of the toes and foot toward the ground, like when a person stands on his tippy-toes.

◆ **Eversion versus inversion.** These terms describe movements of the foot. Eversion occurs when the sole of the foot is moved outward, like when you try to stand on the insides of your feet. Inversion is movement of the sole inward, like when you try

standing on the outsides of your feet. You can probably picture this in your head, so I'll skip the illustration of this one.

Whew! Just like learning a foreign language isn't it? Well, in a very real way, it is. Because our bodies and what we can do with them is so complex, there needed to be a uniform language that all health-care people understood and used in order to communicate clearly with one another. Like any language, however, if you don't use it, you'll lose it; so as we go through this book, you can be sure you will get practice with each of these terms.

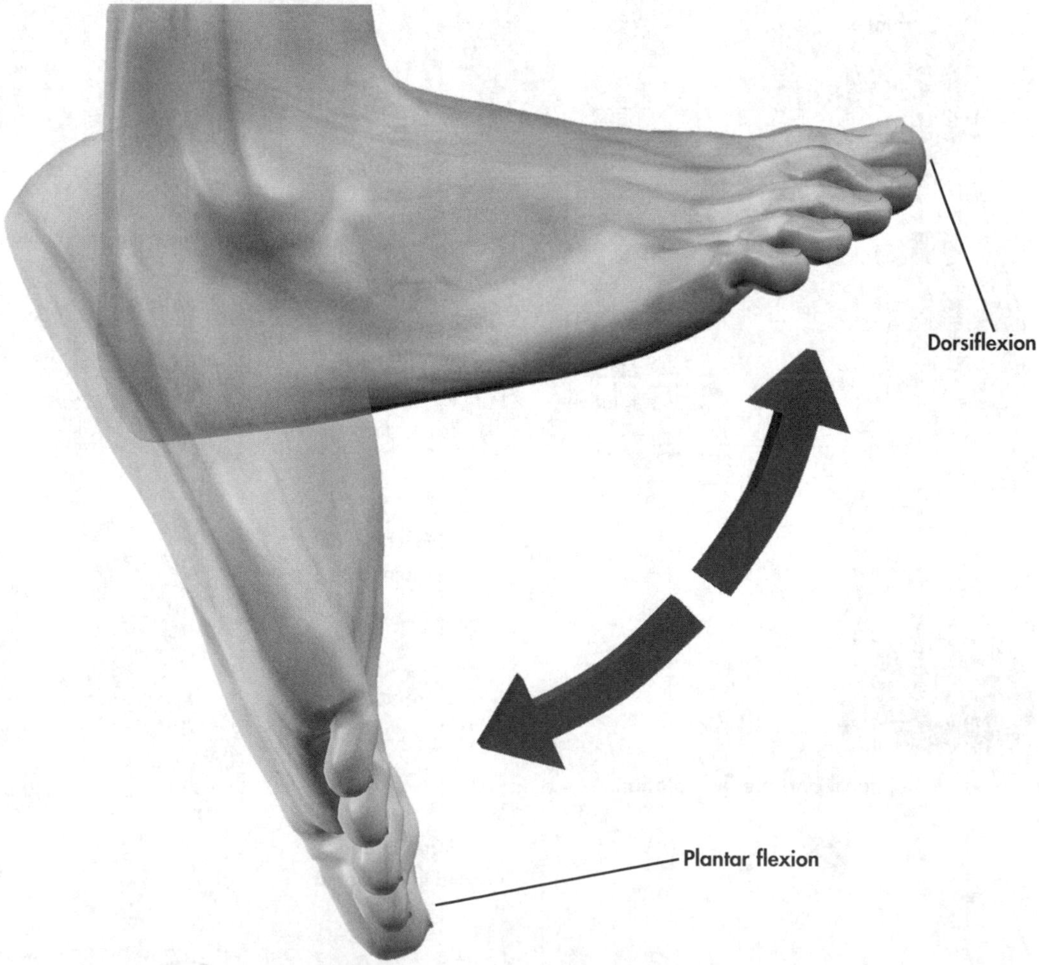

Dorsiflexion

Plantar flexion

Dorsiflexion and plantar flexion.

Foot Notes

The movements described in this section are caused by the action (contraction) of muscles. In upcoming chapters, I describe the muscles or groups of muscles that are responsible for these various movements.

The Least You Need to Know

◆ All descriptions of the body and body parts are made in reference to the anatomical position.

◆ Anatomical planes are imaginary lines that divide the body into different parts.

◆ Anatomical terms of relationship, position, and movement are used in opposites to help communicate information clearly and precisely.

In This Chapter

- ◆ All about the spine
- ◆ Different kinds of vertebrae
- ◆ The spine's joints and ligaments
- ◆ The back and its movements and muscles
- ◆ The spinal cord, spinal nerves, and their coverings

  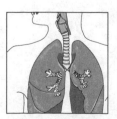 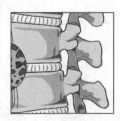 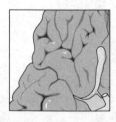 

Chapter **15**

Rear View: The Back

Although the backside is not the part of the anatomy that people spend much time making attractive or that is particularly exposed to the public, it, nonetheless, is a very important region of the body. Back pain, especially low back pain, affects up to 90 percent of Americans at some time in their life. In fact, back pain is one of the most common reasons for seeking medical care. Next to the common cold, back pain is the most frequent cause of lost work days.

In this chapter, you will learn about the different regions of the spine and features of vertebrae in those regions. You'll also learn about the ligaments that hold the vertebral bones together and the joints that enable movement, though limited, between bones. There are also many muscles that attach along the length of the spine that are important for holding a person up, but are very complex in their arrangement. I won't spend much time on these.

You will learn about the spinal cord and its connective tissue covering, the meninges, which is protected by the vertebral column, as well as about the spinal nerves that leave the spinal cord and canal through holes between vertebrae. Finally, we begin the process of integrating concepts and details from Parts 1 and 2, and describe some common injuries and disorders that affect the back.

The Spine: Backbone of the Human Body (Literally)

The vertebral column (also known as the spinal column, spine, or backbone) is a semi-rigid structure consisting of a block-like series of bones, called *vertebrae*, connected by discs, ligaments, joints, and muscles. The vertebral column is made up of 33 vertebrae and divided into five regions: cervical (7 bones), thoracic (12 bones), lumbar (5 bones), sacral (5 sacral vertebrae fused into a single bone, the *sacrum*), coccygeal (4 coccygeal vertebrae fused into a single bone—the coccyx, or "tailbone").

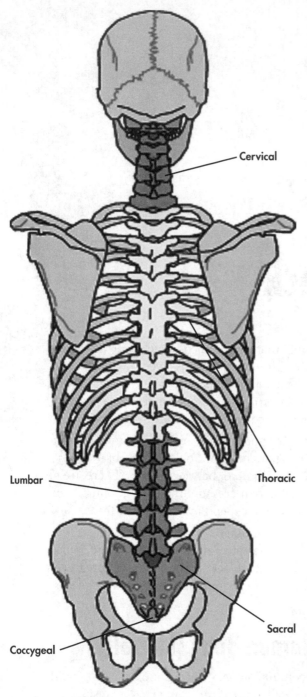

Cervical

Lumbar

Thoracic

Coccygeal

Sacral

The five regions of the spine.

Nice Curves! Normal Curves of the Spine

The normal adult spine has a graceful set of forward and backward curves—one each in the cervical, thoracic, lumbar, and sacral regions. This design is an adaptation to humankind's ability to walk upright on two legs. This spring-like configuration helps cushion and balance the loads a person places on the spine and prevents jarring forces to the brain that would otherwise happen if the spine was straight.

Some of these curves (thoracic and sacral) are present at birth. These are called *primary curves*. The open part of these curves faces toward the front of the body.

The cervical and lumbar curves develop after birth as *secondary curves*. The open part of these curves faces toward the back. Secondary curves form when the infant begins to hold up her own head (the cervical curve) and when she begins to sit up and walk (the lumbar curve). These secondary curves are adaptations in the vertebral column to allow the body's weight to be properly distributed along the spine so that minimal muscular energy is needed to maintain balance.

Throw Me a Curve! Abnormal Curves

Significant changes may occur in these curves, which doctors can observe visibly or by x-ray. Some common abnormal curves and examples include the following:

◆ **Kyphosis.** This is an abnormal or exaggerated thoracic curvature that in severe cases produces a "hump-back" deformity. The "dowager's hump" deformity seen in some elderly women with osteoporosis (see Chapter 4) is an example of a kyphotic curvature. This occurs when weakened thoracic vertebrae develop crush or compression fractures. Changes in the shape of these vertebrae can lead to a stooped, bent-forward posture. (Quasimodo's hunchback was a kyphosis, but he was described as born that way.)

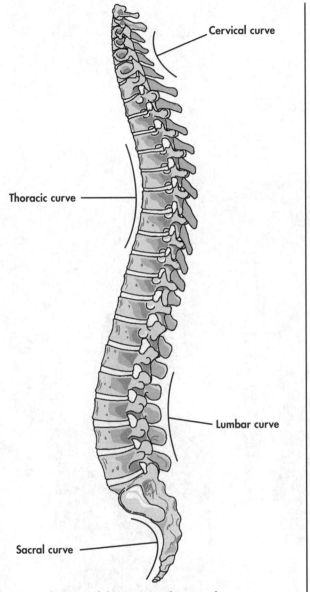

Cervical curve

Thoracic curve

Lumbar curve

Sacral curve

Side view of the spine with normal curves.

◆ **Lordosis.**   This is an abnormal or exaggerated lumbar curve that produces a "swayback" deformity. How might this occur? Ordinarily, the line of gravity for one's body weight passes near or through the front of the lumbar region of the spine. A forward shift in this line of gravity, as happens when extra poundage is added to our midsection, results in a compensatory increase in the lumbar curvature to counter this added weight. This mechanical correction requires increased activity of the lower back muscles. Lordosis also occurs in the late stages of pregnancy, when the weight of the growing fetus causes a temporary change in mom's center of gravity, leading to a compensatory change in the lower back. In both of these examples, the strain placed on the muscles may be a cause of low back pain.

◆ **Scoliosis.**   This is an abnormal side curvature of the spine. These curvatures are typically painless and their cause is often unknown, although it could be due to muscle imbalance or abnormal development of a vertebra. These curvatures are often diagnosed in late childhood or early adolescence (10 to 12 years of age) during a routine screening of the back. A simple forward-bending test may reveal that the ribcage is higher on one side than the other. Also one shoulder and hip might be slightly higher than the opposite side.

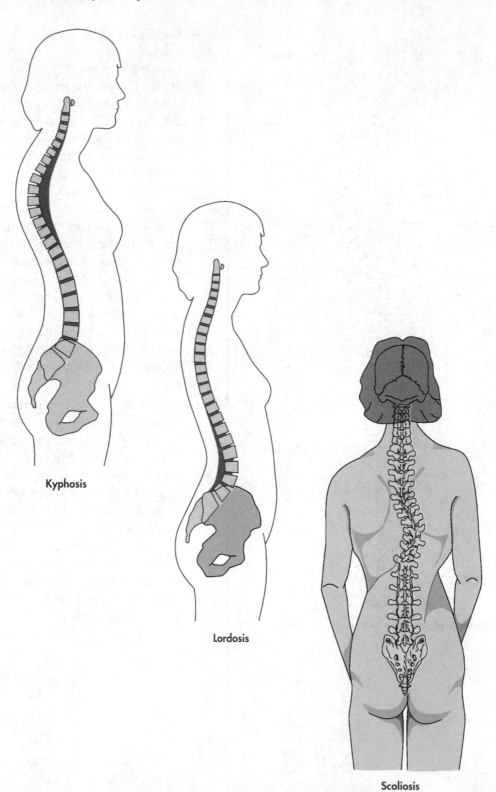

Kyphosis

Lordosis

Scoliosis

Abnormal curves: kyphosis, lordosis, and scoliosis.

Are You from Around These Parts? Parts of a Typical Vertebra

Before you learn about features characteristic to a particular region of the vertebral column, you should know about the typical parts of most vertebrae.

A typical vertebra consists of the following parts:

◆ Body

◆ Vertebral arch

◆ Vertebral foramen

◆ Set of processes

◆ Set of vertebral notches

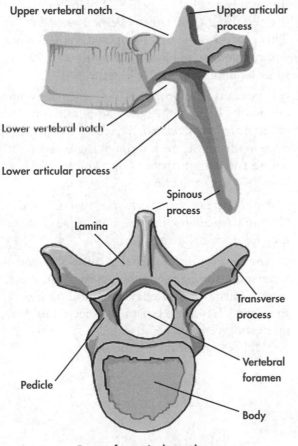

Parts of a typical vertebra.

The *bodies* of vertebrae are their principal weight-bearing parts. They become increasingly larger from the cervical region to the lumbar region. Between the bodies are *intervertebral discs* made of fibrous cartilage (the sites of symphysis-type joints; see Chapter 4).

The bodies attach to the *vertebral arch*, which consists of a pair of *pedicles* (Latin for "little feet") that attach to the body, and a pair of *laminae* that unite with each other in the middle to complete a bony roof. Between the body and vertebral arch is a hole called the *vertebral foramen*. Collectively, this series of *foramina* (the plural form of *foramen*) forms the vertebral canal, which contains and protects the spinal cord and its coverings.

Body Language

Vertebral bodies are the main weight-bearing parts of the vertebrae.

Intervertebral discs are located between vertebrae and are made of fibrous cartilage. The **vertebral arch** is made up of **pedicles** and **laminae**, forming a hole called the **vertebral foramen**, which houses the spinal cord.

Several bony parts (called *processes*) are associated with vertebrae. Most of these act as levers through which muscles move individual vertebrae. These processes are the following:

◆ **Spinous process.** The spinous process is a single process located where the two laminae meet. It sticks out behind at/along the midline. These are the series of "bumps" you can feel when you run a finger down someone's back (or see peeking out the backless gown of your typical twenty-something female celebrity).

◆ **Transverse processes.** These are a pair of processes at the junction between a pedicle and a lamina. They stick outward from this point.

◆ **Articular processes.** These processes, which have upper (superior) and lower (inferior) versions, stick out upward or downward from each pedicle. Articular processes contain shallow depressions called *facets* for articulations between adjoining vertebrae. These joints are synovial joints (see Chapter 4).

◆ **Vertebral notches.** On the respective upper (superior) and lower (inferior) surfaces of the pedicles are vertebral notches. By placing two vertebrae on top of one other, these notches form a hole, called the *intervertebral foramen*. A single spinal nerve leaving the spinal cord exits the vertebral canal through an intervertebral foramen.

What Region Are You From? Regional Characteristics of Vertebrae

Okay, so now that you have the bare-bones basics down on vertebrae, let's step things up a bit and get ready to identify a bone from a particular region of the spinal column. Who knows, your child, niece, nephew, or neighbor kid may take you for a *CSI* agent when they hand you a single bone they picked up on a hike in the woods and you can tell them what part of the critter it's from.

Can We "Neck"? Cervical Vertebrae

The bodies of cervical vertebrae are small compared to those in regions lower down the spine because they have a relatively small amount of weight to support (even for people who have "big heads"). However, the vertebral foramen/canal in cervical vertebrae is large and widest in this part of the spine because the spinal cord has its greatest diameter here.

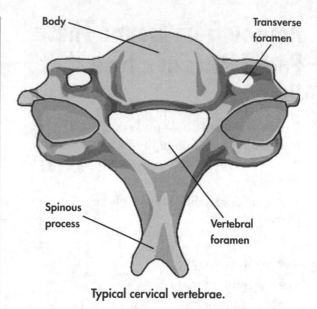

Typical cervical vertebrae.

The spinous processes on cervical vertebrae are typically short and forked, except for one on the vertebra called C7, which is quite long. This is given the special name *vertebra prominens*. This process is the first bump you can feel when you run your finger down the midline of the spine beginning at the back of the skull.

The transverse processes of cervical vertebrae are unique because they have holes in them, called *transverse foramina*. Vertebral arteries come up from the neck through these holes before entering the base of the skull to supply blood to part of the brain.

Among the cervical vertebrae, two (C1 and C2) are particularly unique; so much so that they have special names.

The first cervical vertebra is called the *atlas*. Remember the Greek mythological giant, Atlas, whom Zeus forced to carry the world on his shoulders? His name became connected to this interesting vertebra because it also supports something big and round: the skull.

The atlas looks like a ring of bone. It lacks a body. But it has front and rear arches that connect to the sides of bony enlargements called *lateral masses*. These masses contain the transverse foramina and articular surfaces that attach to the next vertebra and the base of the skull. The synovial joints between the atlas and the skull allow a person to nod his head, "yes."

The *axis*, or second cervical vertebra, has an upward projection from its body, called the *dens*. The dens serves as the body of the atlas. It is held in place behind its front arch by a thick, stout ligament called the *transverse ligament of the atlas*. The dens acts as a pivot to allow rotation between the atlas and axis vertebrae, like when a person shakes her head to say, "no."

Body Language

Vertebra prominens is a long spinous process on the C7 vertebra. It's the first bump you can feel at the bottom of your neck. **Transverse foramina** are the holes in the transverse processes of the neck vertebrae. The **atlas** is the top vertebra in the neck (also called C1). It is oval and has no body. The **axis** is the second neck vertebra (C2). It helps support the atlas and enables you to turn your head from side to side.

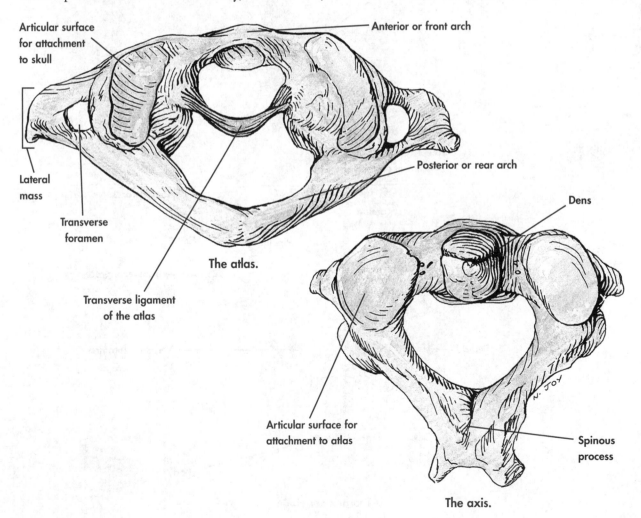

Articular surface for attachment to skull

Anterior or front arch

Lateral mass

Posterior or rear arch

Transverse foramen

The atlas.

Transverse ligament of the atlas

Dens

Articular surface for attachment to atlas

Spinous process

The axis.

Thoracic Vertebrae: This Will Stick to Your Ribs

The thoracic vertebrae have pretty much the same features as a typical vertebra. There are 12 thoracic vertebrae. The bodies of these vertebrae are heart-shaped and larger than cervical vertebrae. They have a set of small, shallow depressions along each upper and lower edge, called *costal (rib) demifacets*, where the ribs attach. In addition, most transverse processes also have facets for articulation with ribs. These are sites of costovertebral joints. The spinous processes of thoracic vertebrae are long and slender and directed downward. They overlap each other like shingles on a roof.

Lumbar Vertebrae (a Pain in the Lower Back)

Because more of a person is supported in the lumbar region, it is understandable that everything about these vertebrae is bigger. They have massive bodies. Their spinous processes are rectangular and directed horizontally backward. These do not overlap each other, a detail that doctors take advantage of during a spinal-tap procedure.

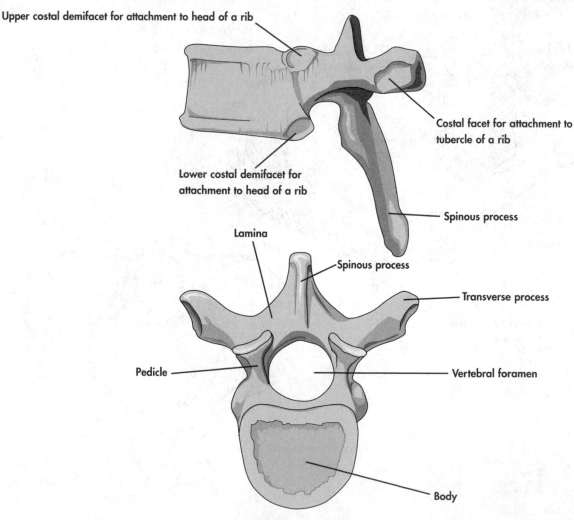

Upper costal demifacet for attachment to head of a rib

Costal facet for attachment to tubercle of a rib

Lower costal demifacet for attachment to head of a rib

Spinous process

Lamina

Spinous process

Transverse process

Pedicle

Vertebral foramen

Body

Thoracic vertebrae.

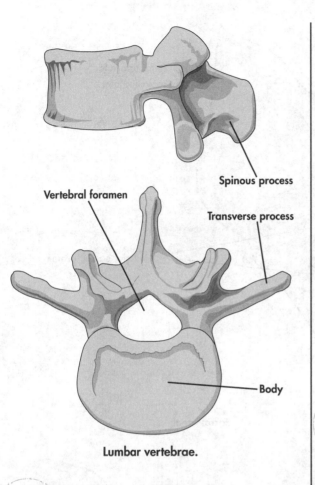

Spinous process

Vertebral foramen

Transverse process

Body

Lumbar vertebrae.

Back to Back, Sacroiliac: The Sacrum

The adult sacrum is a single triangular bone. Its broad base articulates with the body of the fifth lumbar vertebra (called L5). Its narrow apex articulates with the coccyx (tailbone). Along each side of the sacrum are articular surfaces for attachment to the hip bone at the *sacroiliac joint.* The sacrum contains a bony canal, called the *sacral canal,* which is connected to the vertebral canal above it. Spinal nerves descend into the sacral canal and contain the *dural sac* full of nerve roots.

On both front and rear surfaces of the sacrum are four pairs of holes for passage of the sacral nerves to peripheral tissues. The sacral canal ends as an opening called the *sacral hiatus.* The delicate, filament-like end of the spinal cord passes through this hiatus before attaching to the tailbone.

Body Language

The **sacroiliac joint** is the joint where the sacrum joins the hip bone. The **sacral canal** is a hole through the sacrum that aligns with the vertebral canal and contains the lower end of the **dural sac** with its sacral and coccygeal nerve roots. The **sacral hiatus** is the opening at the end of the sacral canal.

Bodily Malfunctions

Back pain is not a specific disorder, but rather a symptom of many potential sources that have to be unraveled by a doctor. Common causes of lower back pain include strains to back muscles or ligaments from a sports injury, automobile accidents, falls, or lifting something too heavy; disc herniation and nerve root compression; and osteoarthritis.

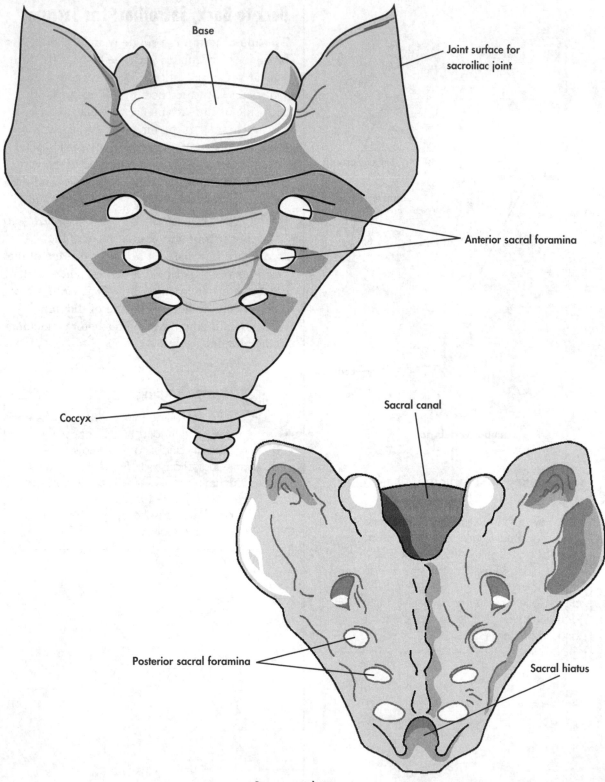

Sacrum and coccyx.

Coccyx: Not a Prehistoric Bird

The coccyx is the "tail end" of the spine. It consists of three to five (usually four) small vertebrae that have lost their resemblance to vertebrae. Rather, they appear as segments of solid bone connected to one another by small amounts of fibrocartilaginous tissue.

The coccyx is a triangular structure. It is attached at its base to the sacrum by a fibro-cartilaginous joint, which permits a limited amount of movement. The coccyx gets its name from a Greek word meaning "cuckoo" because it has the shape of a cuckoo's beak. (This also describes most people's mental state when they have taken a heavy fall on their bottom side and have severely bruised or broken their coccyx. This is a very painful injury!)

Action at the Vertebral Joint

There are two types of joints in the vertebral column: symphyseal and synovial (see Chapter 4 for more on these). Synovial joints occur at the articular processes and at costal facets where ribs attach.

The bodies of vertebrae are held tightly to one another by *intervertebral discs* at symphysis joints. These are shock-absorber devices that "give" in response to compressive forces on the spine. You can think of individual discs as a jelly donut. They consist of a tough outer margin of dense, fibrous connective tissue (the *anulus fibrosus*) and a soft, jelly-like inner core called the *nucleus pulposus*. Although the extent of the movement between individual vertebrae is small, considerable movement is allowed over the length of the spine.

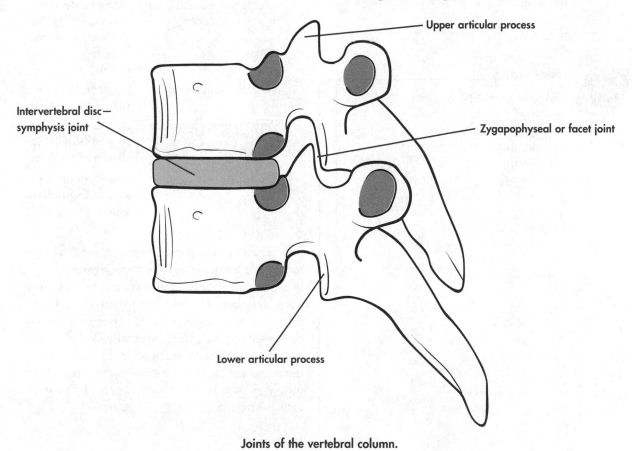

Upper articular process

Intervertebral disc— symphysis joint

Zygapophyseal or facet joint

Lower articular process

Joints of the vertebral column.

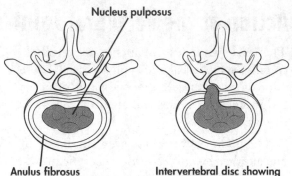

Nucleus pulposus

Anulus fibrosus

Intervertebral disc showing
herniation of nucleus pulposus

The two parts of an intervertebral disc.

Bodily Malfunctions

It is the softer, gelatinous material of the *nucleus pulposus* that protrudes during a disc herniation (a "slipped disc") to cause pressure on the spinal cord or a spinal nerve. This causes pain or numbness.

Foot Notes

Approximately 25 percent of the length of the spine is in the discs. Young people are slightly taller when they hop out of bed in the morning than when they jump into bed at night. Why? The discs become hydrated during rest. During the course of a day on your feet, some of this water gets squeezed out and you lose a tiny bit of height.

Ligaments of the Vertebral Column

A variety of ligaments bind various parts of each vertebra together to hold it as a single unit. I list their names and bony attachments in the following table and identify them in the following figure.

Ligaments of the Vertebral Column

Name of Ligament	Attachment Sites
Anterior longitudinal ligament	Along anterior surfaces of vertebral bodies and intervertebral discs from skull to sacrum
Posterior longitudinal ligament surfaces	Along posterior of vertebral bodies and intervertebral discs from skull to sacrum
Ligamentum flavum	Between laminae of adjacent vertebrae
Interspinous ligament	Between spinous processes of adjacent vertebrae
Supraspinous ligament	Connects tips of adjacent spinous processes

The anterior and posterior longitudinal ligaments are particularly important ligaments because they "splint" the vertebral bodies and discs together. The anterior longitudinal ligament prevents hyperextension (bending too far backward) of the spine, especially in the neck region. Conversely, the posterior longitudinal ligament serves to resist hyperflexion (bending too far forward) of the spine. It also helps prevent herniation of the discs.

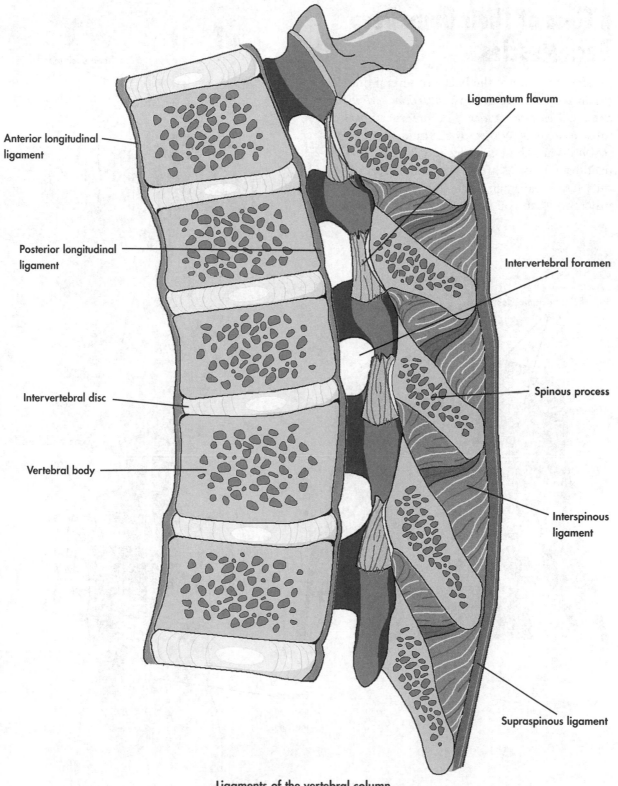

Anterior longitudinal ligament

Posterior longitudinal ligament

Intervertebral disc

Vertebral body

Ligamentum flavum

Intervertebral foramen

Spinous process

Interspinous ligament

Supraspinous ligament

Ligaments of the vertebral column.

A Class of Their Own: Deep Back Muscles

Beneath the skin of the back are several muscles organized into two groups: *superficial back muscles* and *deep back muscles*. The superficial back muscles actually control the shoulder and arm, so I'll discuss these muscles in Chapter 16. The deep back muscles are a very complex set of muscles. A thorough discussion of them is just way more than you probably want to know.

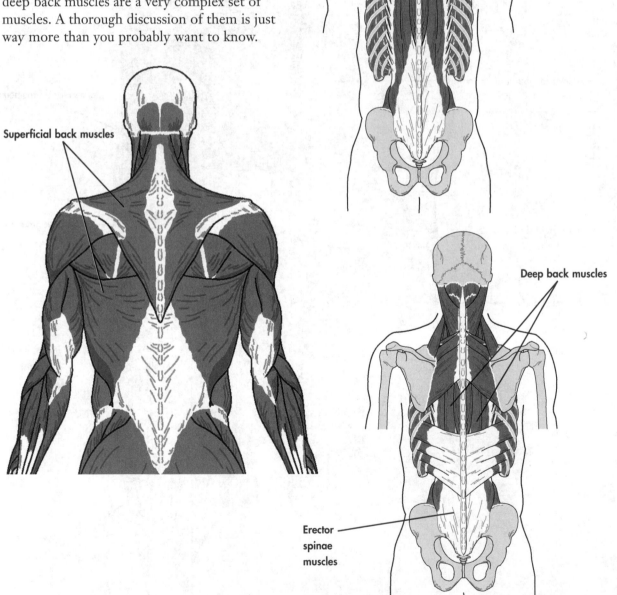

Muscles of the back (superficial and deep).

But let me summarize the chief features of these muscles. The deep back muscles consist of overlapping sets of muscles, which are arranged in longitudinal columns next to the spine. These muscles, collectively, attach along the length of the spine from the skull down to the sacrum. When they contract, they operate on the joints of the vertebral column to extend it. They also enable a person to bend their back from side to side.

The *erector spinae* group is a particularly large and powerful group of muscles in the lower back that helps maintain posture and balance. These muscles are also important in controlling flexion of the vertebral column produced by other muscles and in maintaining the lumbar curvature. That is why they need to be kept strong and in shape by regular exercise.

Body Language

The **erector spinae** is a powerful group of muscles that helps keep your spine erect.

The deep back muscles receive their nerve supply from branches of individual spinal nerves called dorsal rami (literally, "back branch") that were described in Chapter 6.

Spinal Cord, Spinal Nerves, and Meninges

The spine contains and protects the spinal cord and its protective coverings, the meninges, within its vertebral or spinal canal. The spinal cord and meninges are enclosed within fat that fills the space between the meninges and the walls of the spinal canal.

During early fetal development, all spinal nerves exit their respective intervertebral foramina horizontally because the spine and spinal cord are the same length. With subsequent development of the spine outpacing the growth of the spinal cord, the spinal cord takes a position higher up within the spinal canal. As a result, the lower spinal nerve roots (lumbar and sacral) are long and must descend farther in the canal before exiting their foramina. This collection of long nerve roots resembles a horse's tail, so it's often referred to as the *cauda equina*.

Chapter 5 has a more detailed discussion of the spinal cord, the spinal nerves, and the meninges.

Foot Notes

Samples of CSF (used for clinical testing) are routinely obtained by a procedure known as a lumbar puncture or spinal tap. Because the spinal cord ends at the L2 level, a needle can be inserted through the skin and tissues below that level and into the subarachnoid space to collect a sample for analysis—without damaging the spinal cord. This is also the spot where the needle is inserted during labor (called an epidural) to numb the mother from the waist down.

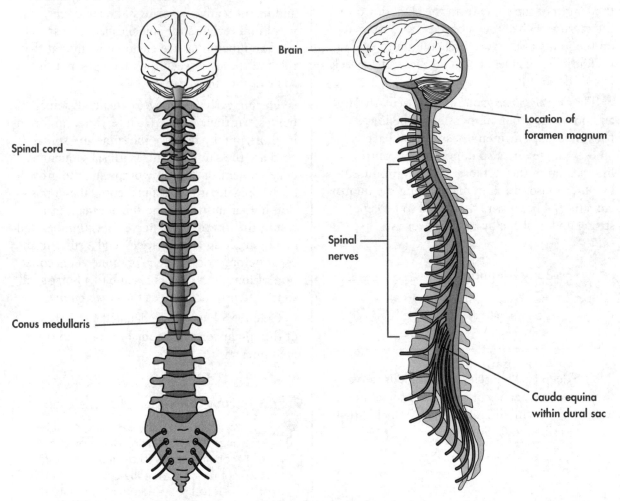

Brain

Spinal cord

Conus medullaris

Location of
foramen magnum

Spinal
nerves

Cauda equina
within dural sac

Brain, spinal cord, and spinal nerves.

The Least You Need to Know

◆ The vertebral column is a semi-rigid structure of vertebrae, discs, ligaments, and joints that give the spine its great flexibility and movement.

◆ The vertebral column contains and protects the spinal cord and its coverings (meninges).

◆ The anatomy of the back is very complex and explains why so many people suffer back pain and why it's so difficult to diagnose and treat.

In This Chapter

- ◆ The organization of the upper limb and its components

- ◆ Features of bones and muscle attachments

- ◆ The actions of muscles within the compartments of the upper limb

- ◆ The nerve and blood supply to the upper limb

- ◆ Common injuries to the upper limb

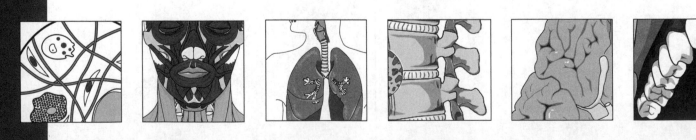

Chapter 16

The Upper Limb: From the Waist Up

The upper limb is made up of the shoulder, arm, forearm, and hand. Probably more than any other region of the body, it defines our personal independence, productivity, and sense of well-being. So much of what we do or want to do is, "hands on." This pretty much defines its function. The upper limb is designed to allow us to position our hands to carry out the many activities that fill our lives.

How is this accomplished? Through a marvelous system of bones and joints that provide great mobility and maneuverability of the limb to an object and then the ability to manipulate that object. This chapter will teach you the basic content and organization of the upper limb and about common types of musculoskeletal injuries to help you understand how it works in real life.

The bones and joints in each of these areas are described first, followed by a description of the muscles associated with each area. Be sure to review the anatomical position terms in Chapter 14 before diving into this chapter. Details describing features of synovial joints are described in Chapter 4.

Bones and Joints of the Upper Limb

This section covers the bones and joints of the shoulder, the arm, the forearm, the elbow, the wrist, and the hand. The muscles of these areas are described in a later section in this chapter.

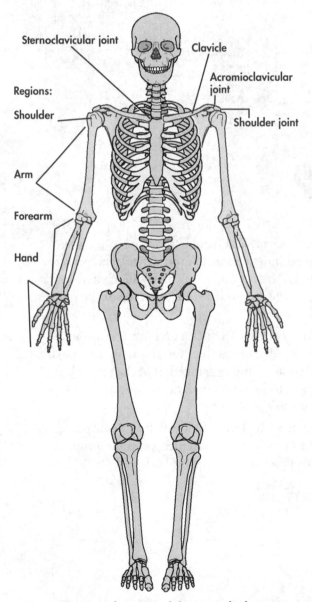

Bones and regions of the upper limb.

Labels: Sternoclavicular joint, Clavicle, Acromioclavicular joint, Shoulder joint, Regions:, Shoulder, Arm, Forearm, Hand

Bones of the Shoulder Region

The shoulder is where the upper limb connects to the trunk of the body. The bones of the shoulder region include the *clavicle*, the *scapula*, and the upper (proximal) part of the *humerus* (not a laughing matter!). There are important features on these bones that you need to know because they relate to where muscles or ligaments attach and will help you understand muscle and joint actions.

> **Body Language**
>
> The **clavicle** is the collarbone. The **scapula** is the shoulder blade. The **humerus** is the bone of the arm.

The *clavicle* is an S-shaped bone located just beneath the skin. You can feel it between the sternum and the tip of the shoulder. It is mostly cylindrical in shape, but expands at its two ends. Its medial end attaches to the *sternum* (also called the breastbone). Its flat lateral end attaches to the *acromion process* of the scapula (or shoulder blade). The clavicle is a very important bone of the upper limb because it serves as a strut to keep the limb away from the body so that it can move about freely. The clavicle is the most commonly fractured bone.

The scapula is a thin, triangular-shape bone that is sometimes referred to as the shoulder blade. It is located on the upper part of the back. Each of the two scapulae has three borders (superior, medial or vertebral, and lateral or axillary) and three angles (superior, lateral, inferior). It also has three shallow surface depressions called *fossae*, where muscles attach.

The bony ridge that runs across (medial to lateral) the back side of the scapula is called the spine of the scapula. It ends laterally in a flat, bony projection called the *acromion*. This attaches to the clavicle and forms the point of the shoulder.

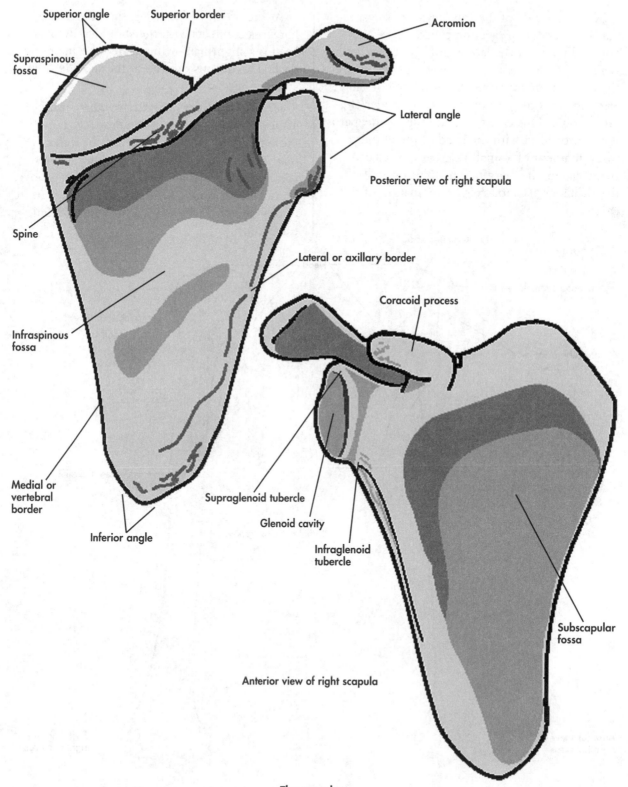

Superior angle

Superior border

Acromion

Supraspinous fossa

Lateral angle

Posterior view of right scapula

Spine

Lateral or axillary border

Coracoid process

Infraspinous fossa

Medial or vertebral border

Supraglenoid tubercle

Glenoid cavity

Inferior angle

Infraglenoid tubercle

Subscapular fossa

Anterior view of right scapula

The scapula.

Along the side of the top border of the scapula is a beak-like projection called the *coracoid process*. Finally, at the lateral angle of the scapula is a shallow depression, the *glenoid cavity*, where the head of the humerus (the top bone of the arm) attaches to form the shoulder, or glenohumeral joint. This cavity is made slightly deeper for improved fit with the head of the humerus by attachment of a small ridge of fibrous cartilage around its margin. The joint capsule of the shoulder also attaches to the margin of this cavity.

At the proximal humerus, the head of the humerus connects with the shoulder. Below the head is a slight narrowing, called the anatomic neck. The fibrous capsule of the shoulder joint attaches here.

Greater and lesser tubercles are where muscles associated with the rotator cuff attach. Between these tubercles is a groove called the *intertubercular* or *bicipital groove*. A tendon of the *biceps brachii* (long head) normally rests in this groove. It attaches to a bony point on the upper margin of the glenoid cavity (called the supraglenoid tubercle).

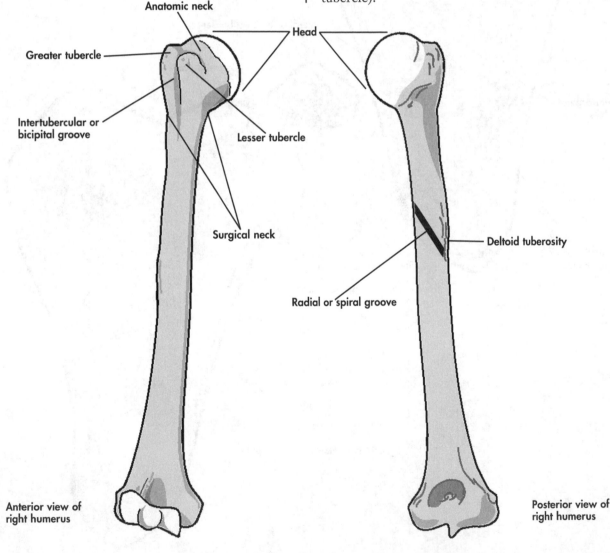

Anatomic neck

Head

Greater tubercle

Intertubercular or bicipital groove

Lesser tubercle

Surgical neck

Deltoid tuberosity

Radial or spiral groove

Anterior view of right humerus

Posterior view of right humerus

The humerus.

Below the tubercles is the surgical neck of the humerus. This is another common site of fracture. The axillary nerve and posterior circumflex humeral artery course behind the surgical neck. These are particularly vulnerable to injury during a fracture at this site. From here, the humerus narrows into a cylindrical shaft.

On the back side of the humerus about mid-length is a diagonal, shallow depression called the *radial* or *spiral groove*. The radial nerve and the deep brachial artery follow in this groove (more on these later). Finally, about halfway down the side of the humerus is a roughened area called the *deltoid tuberosity*. This is where the deltoid muscle attaches.

Joints of the Shoulder Region

Four joints are associated with the shoulder region:

◆ **Sternoclavicular joint.** A synovial joint that pivots during movement of the shoulder.

◆ **Acromioclavicular joint.** A synovial joint that glides during movement of the shoulder.

◆ **Glenohumeral joint.** A ball-and-socket synovial joint that allows the arm to flex, extend, adduct, abduct, medially or internally rotate, laterally or externally rotate, and circumduct or move in a 360-degree arc.

◆ **Scapulothoracic joint.** A "fake" joint that does not have a bony connection of any kind. It is actually a space between the scapula and the surface of the rib cage where the scapula slides forward (as to extend one's reach) and backward and rotates upward (as to raise the arm) and downward against the rib cage.

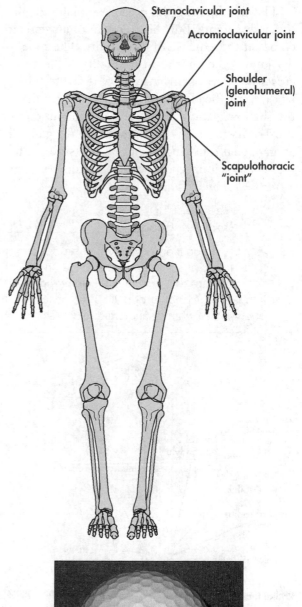

The joints of the shoulder region.

The shoulder has the greatest mobility of all the body's joints. But this freedom of movement comes at the expense of its stability. The shoulder joint is the joint most frequently subject to *dislocation*. Part of this instability is due to the fact that the head of the humerus is larger than the shallow glenoid cavity, and only about one third of its surface is in contact with the glenoid at any given time. It's like a golf ball sitting on a tee. When a joint is dislocated, doctors perform a *joint reduction* to fix it.

Body Language

A **dislocation** occurs when the opposing joint surfaces are no longer in contact with one another. **Joint reduction** is the procedure of putting the bones back to their normal anatomic relationship.

The shoulder's fibrous joint capsule is reinforced by ligaments that blend into the capsule. But its greatest strength and stability is provided by the tendons of the rotator cuff muscles—supraspinatus, infraspinatus, teres minor, and subscapularis—which tightly fuse to the joint capsule and hold the head snugly in the joint.

Bones of the Arm and Forearm

When most people talk about the arm, they are usually referring to everything in the upper limb except, perhaps, the hand. When an anatomist considers the arm, it is a specific part of the upper limb: that is, the part between the shoulder and the elbow. The forearm, then, is the part of the upper limb between the elbow and the wrist, and the hand is the end of the upper limb.

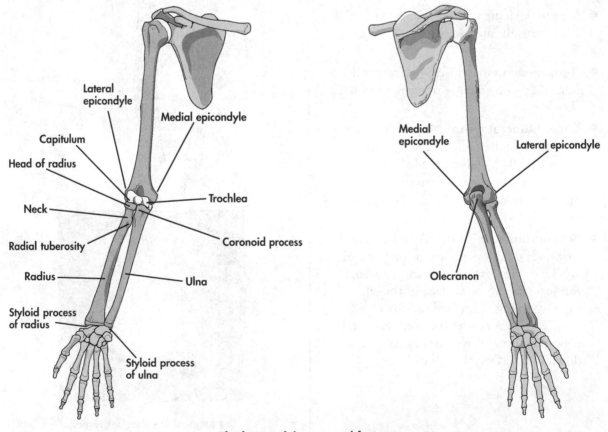

The bones of the arm and forearm.

The bone of the arm is the humerus. The upper part of the humerus attaches to the scapula at the shoulder joint. The humerus has a cylindrical shaft that widens at its lower end. Here, it attaches to the two bones of the forearm, the radius and the ulna, at the elbow joint.

On the outside surface of the lower end of the humerus is a bony projection that is easily felt. This is the lateral epicondyle. A somewhat larger projection can be felt on the inside surface of the elbow, called the medial epicondyle. On the backside of the medial epicondyle is the ulnar groove, for the ulnar nerve. This is the "funny bone." Several muscles of the forearm attach to these epicondyles.

Below each epicondyle and marking the end of the humerus are a pair of condyles. The lateral of the two condyles is a rounded projection of bone called the capitulum. The head of the radius articulates here. The medial of the two is a "pulley"-like bony projection called the trochlea. This articulates with the ulna.

The two bones in the forearm are the *radius* and the *ulna*. The radius is the lateral of the two bones. At its proximal end is a round, disc-like head. Beneath it is a narrow region called the neck. Below the neck is a small projection of bone, called the *radial tuberosity*, where the tendon of the biceps brachii muscle attaches. The shaft of the radius continues and broadens at the wrist. On the medial side of the distal radius is the *ulnar notch*, for attachment to the head of the ulna.

Body Language

The two bones of the forearm are the **radius** and the **ulna**. They run alongside one another and connect to the elbow joint on one end and the wrist joint on the other.

The proximal part of the ulna has a prominent projection of bone on its back surface called the *olecranon process*. This is the "point" of the elbow. In front of this process is a deep concavity known as the *trochlear notch*. Just below the lip of the trochlear notch is the *coronoid process*, which has a roughened portion of bone for attachment of the brachialis muscle of the arm. Along the lateral margin of the trochlear notch is a shallow depression called the *radial notch* where the head of the radius articulates.

The ulna has a long, narrow shaft that ends at the wrist. A smaller styloid process extends off the posterior surface of the ulna. It is readily seen and felt as a round "bump" through the skin on the backside of the wrist. The shafts of these bones are connected by a fibrous membrane called the *interosseous membrane*, to which some of the forearm muscles attach.

The Elbow and Radioulnar Joints

The elbow joint is primarily a hinge-type joint that allows the forearm to bend (flex) and straighten out (extend). It is unique because it has three different articulations in one joint cavity:

◆ The *humeroradial articulation* between the capitulum of the humerus and the head of the radius.

◆ The *humeroulnar articulation* between the trochlea of the humerus and trochlear notch of the ulna.

◆ The *radioulnar articulation* between the head of the radius and the radial notch of the ulna.

In this latter articulation, the head of the radius is held snugly to the ulna by the ring-like *annular ligament*, which surrounds the radial head and attaches to the ulna.

There are two radioulnar articulations at opposite ends of the forearm bones. The one at the elbow joint is the *proximal radioulnar joint*, whereas the *distal radioulnar joint* is at the wrist. Rotational movements of supination and pronation of the forearm happen at these joints.

Bodily Malfunctions

The head of the radius in a very young child is not as developed as it is in an older person. Sudden upward jerks on a toddler's arm (as in attempting to catch him from falling or while lifting her up by the hand) may cause the head of the radius to slip out of position, resulting in sudden pain around the child's elbow. This partial dislocation is called radial subluxation, or Nursemaid's elbow.

Bones and Joints of the Wrist and Hand

The wrist contains eight irregularly shaped carpal bones arranged in two rows of four (see the following figure for their names). The distal row of carpal bones articulates with the metacarpal bones in the palm of the hand. The metacarpals, in turn, articulate with the bones of the digits, called *phalanges*. The thumb contains two phalanges, and each of the fingers contains three phalanges.

The wrist joint is known as the *radiocarpal joint*. It consists of the broad end of the radius along with a triangular-shape articular disc that attaches between the radius and the ulna. This disc separates the wrist joint from the distal radioulnar joint.

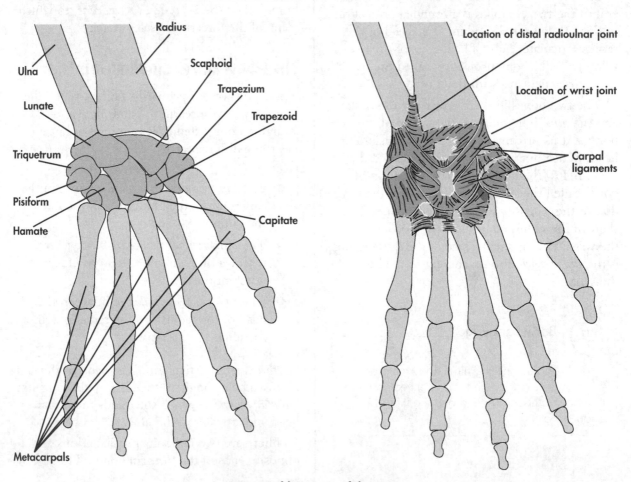

Bones and ligaments of the wrist.

Foot Notes

The arrangement of bones, joints, and ligaments means that forces directed onto the hand and wrist during a fall are first transmitted through the scaphoid and lunate bones to the distal radius, transferred across the interosseous membrane to the ulna, and continued on to the humerus. This explains why the scaphoid bone is the most frequently fractured bone of the wrist, why the distal radius is so frequently fractured, and why the lunate and ulna bones are commonly dislocated when a person falls on an outstretched hand. It also illustrates that the upper limb, unlike the lower limb, is not a weight-bearing structure!

The carpal bones are connected to one another by very short, tough ligaments. They are tightly bunched together and have limited gliding movements between each other. Their joint capsules are reinforced by strong ligaments.

There are several types of joints in the hands:

◆ **Carpometacarpal (CMC) joints.** The joints between the distal row of carpal bones and the metacarpals. The most mobile and important of these is the first CMC joint in the thumb. This joint is nearly 90 degrees out of plane with the other CMC joints. This is important, because it allows the thumb great mobility and the ability to move itself to the other fingers, as when grasping a pencil or pinching some object.

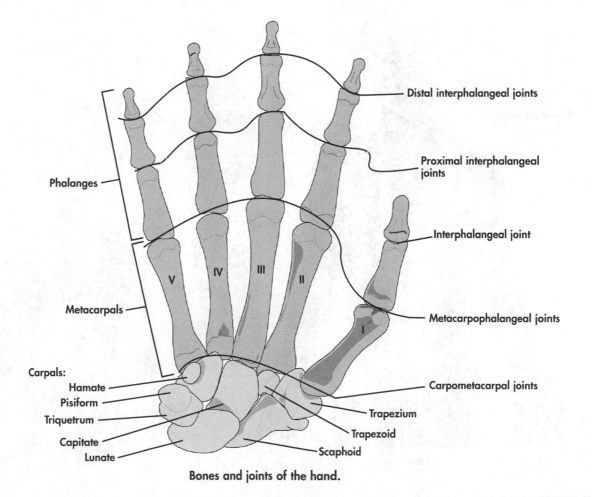

Bones and joints of the hand.

◆ **Metacarpophalangeal (MCP) joints.** Located at the knuckles of the hand. They are the articulations between the metacarpal bones and the proximal phalanges.

◆ **Interphalangeal (IP) joints.** The joints between the phalanges. There is one interphalangeal joint in the thumb and two interphalangeal joints (a proximal IP or PIP joint and a distal IP or DIP joint) in each of the fingers.

Muscles of the Upper Limb

The parts of the upper limb are moved by numerous muscles. These muscles are assembled into functional groups. Those surrounding and moving the shoulder are less numerous but more massive and powerful than those farther down the limb. At the opposite end, in the hand, the muscles are small, numerous, and designed for fine, precision movements of the fingers. The muscles of each of these areas are discussed in the following sections.

Muscles of the Shoulder Region

More than a dozen muscles are involved in controlling movements of the shoulder at the shoulder and scapulothoracic joints. It is easiest to organize these muscles into functional groups: those that move the arm and those that move the scapula. Remember that movement at one of these joints commonly produces movement at the other.

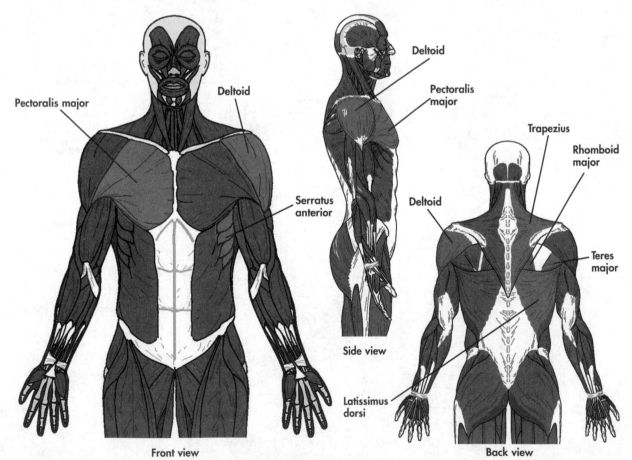

Muscles of the shoulder region.

Some of the most important shoulder muscles include the following:

◆ **Trapezius.** This is the largest muscle that moves the scapula. Because of its size, contraction of different parts of it produces different movements of the scapula. Its upper fibers are used to elevate the scapula as when a person shrugs their shoulders. Its middle fibers can retract the scapula as when squaring one's shoulders, and its lower fibers rotate the scapula upward as when a person reaches to remove something from a top shelf. Raising and retracting the scapula are aided, respectively, by the *levator scapulae* and the *rhomboid* muscles, which are located beneath the trapezius.

◆ **Serratus anterior.** This is an important muscle in protracting the scapula, as when a person wants to increase their forward reach, and in rotating the scapula upward. It also fixes or presses the scapula firmly against the rib cage to allow other muscles to move the arm.

◆ **Deltoid.** This is a very important and powerful muscle of the shoulder joint and it abducts the arm.

◆ **Pectoralis major.** This is a powerful adductor and medial rotator of the arm.

◆ **Latissimus dorsi.** This muscle primarily extends and medially rotates the arm. It is used when a person swims a crawl stroke or chins on a bar.

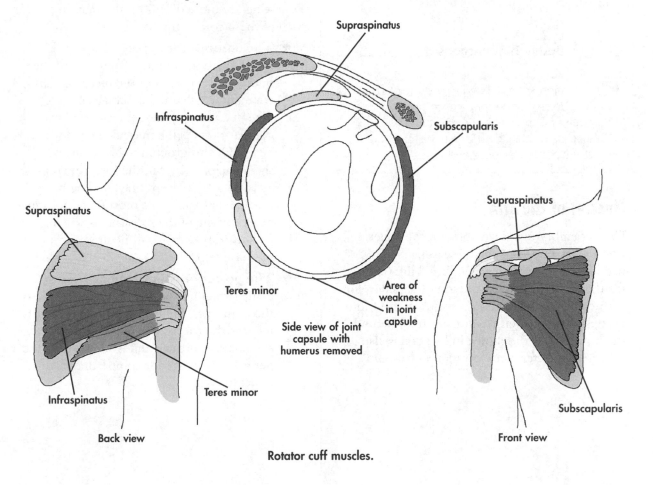

Rotator cuff muscles.

The most important muscles of the shoulder, however, are the muscles that make up the rotator cuff. They help stabilize the shoulder. These muscles are the following:

- **Supraspinatus.** Starts abduction of the arm, which is subsequently taken over by the deltoid.
- **Infraspinatus.** Laterally rotates the arm.
- **Teres minor.** Laterally rotates the arm.
- **Subscapularis.** Medially rotates the arm.

Their tendons fuse with the joint capsule and strengthen it on all sides except its inferior surface. This area is potentially weaker than the other walls. In fact, it is through this portion that greater than 95 percent of shoulder dislocations occur.

Bodily Malfunctions _____

Rotator cuff rupture or tear is a common sports-related or work-related injury for people whose activities require much over-the-head type of work (such as a pitcher). This injury is characterized by shoulder pain and instability.

Muscles of the Arm

During embryonic development, the upper limbs grow out from the body as pairs of paddle-like structures called limb buds (likewise for the lower limbs). As the long bones develop within them, muscle tissue assembles itself into groups of muscle, either in front of or behind the bones. This difference in location is also reflected in differences in function between these muscle groups. Muscles located in front of the bones of the upper limb are described as being in the *anterior compartment* of the limb. Most of these muscles flex the forearm, wrist, and hand. For this reason, the anterior compartment is also called the *flexor compartment*. Opposite to this, muscles located behind the long bones of the upper limb are described as being in the *posterior compartment*. Most of these muscles extend the forearm, wrist, and hand, so this compartment is also called the *extensor compartment*. This pattern of development makes it easier to group the muscles of the limbs and their functions. Also most compartments are supplied by a single peripheral nerve.

The muscles of the arm are primarily responsible for flexing or extending the elbow. Following are more details on the muscles in each of these compartments:

- **The anterior (or flexor) compartment.** Located on the front side of the humerus. It consists of the biceps brachii, coracobrachialis, and brachialis muscles. The nerve supply to the muscles of this compartment is the musculocutaneous nerve. Of this group, the brachialis is the principal flexor of the forearm and is assisted by the biceps brachii. The biceps brachii also acts as the most powerful supinator muscle of the forearm, as when a person tries to turn a difficult screw into a piece of wood.
- **The posterior compartment.** This compartment is located on the back side of the humerus. The muscle of this compartment is the triceps brachii. The triceps extends the forearm and is supplied by the nerve of the posterior compartment, the radial nerve.

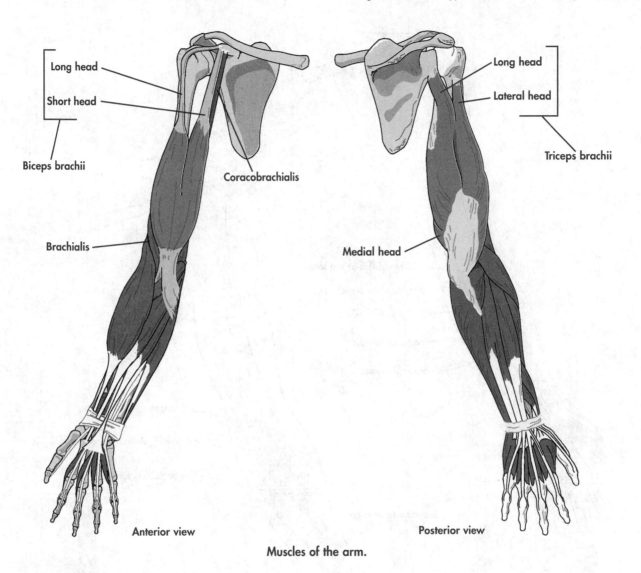

Long head
Short head
Biceps brachii
Brachialis
Coracobrachialis
Anterior view

Long head
Lateral head
Triceps brachii
Medial head
Posterior view

Muscles of the arm.

Muscles of the Forearm

The muscles of the forearm, like those of the arm, are grouped according to their location on the front side or back side of the bones of the forearm. The muscles of the anterior compartment of the forearm (also called the flexor compartment) primarily flex the wrist and the fingers. The muscles of the posterior compartment primarily extend the wrist and fingers, and thus are also called the extensor compartment of the forearm. Following is a brief discussion of the key elements of these two compartments.

Most of the muscles of the flexor compartment attach to the humerus' medial epicondyle, while many of the extensor compartment muscles attach to the humerus' lateral epicondyle. The tendons of these muscles are long and narrow, in order to minimize bulk where they cross the wrist. Several of the latter can be readily seen through the skin on the back of the hand. These tendons are held in place at the back of the wrist by a thickening of tough connective tissue called the *extensor retinaculum*. This prevents the tendons from "bowstringing" when the fingers are extended.

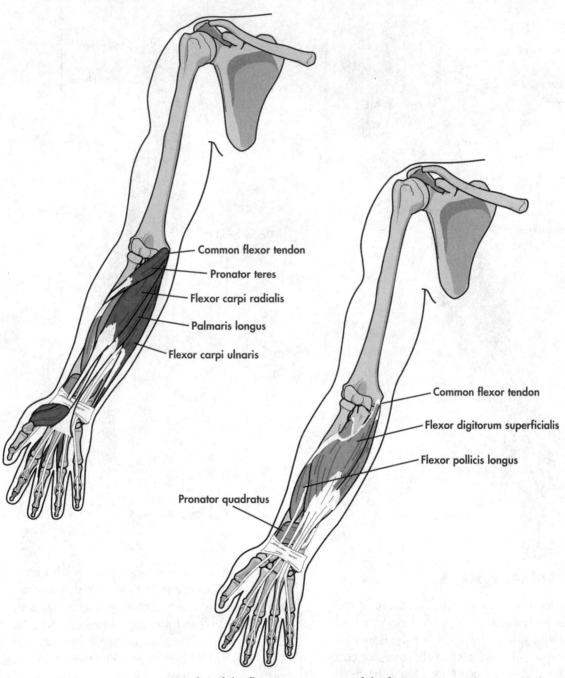

Common flexor tendon
Pronator teres
Flexor carpi radialis
Palmaris longus
Flexor carpi ulnaris

Common flexor tendon
Flexor digitorum superficialis
Flexor pollicis longus
Pronator quadratus

Muscles of the flexor compartment of the forearm.

The names and locations of the muscles in the flexor and extensor compartments are shown in the following figures.

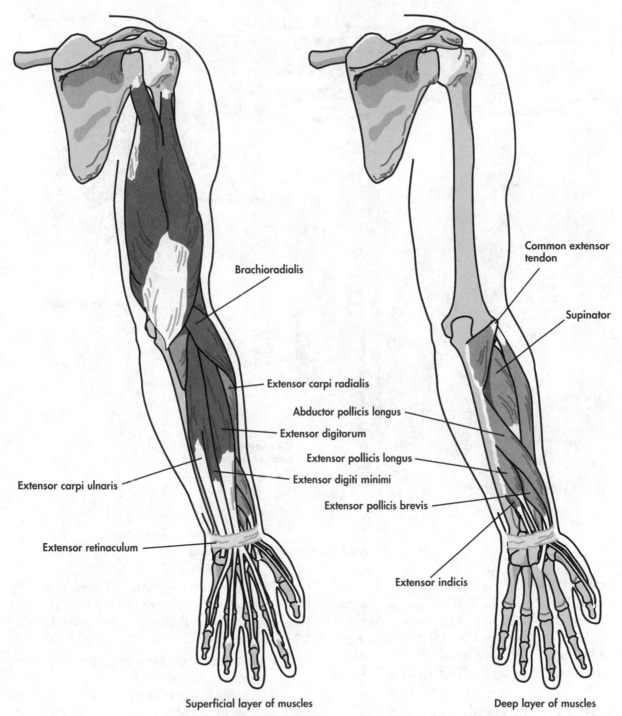

Brachioradialis

Extensor carpi radialis

Extensor digitorum

Extensor carpi ulnaris

Extensor digiti minimi

Extensor retinaculum

Common extensor
tendon

Supinator

Abductor pollicis longus

Extensor pollicis longus

Extensor pollicis brevis

Extensor indicis

Superficial layer of muscles

Deep layer of muscles

Muscles of the extensor compartment of the forearm.

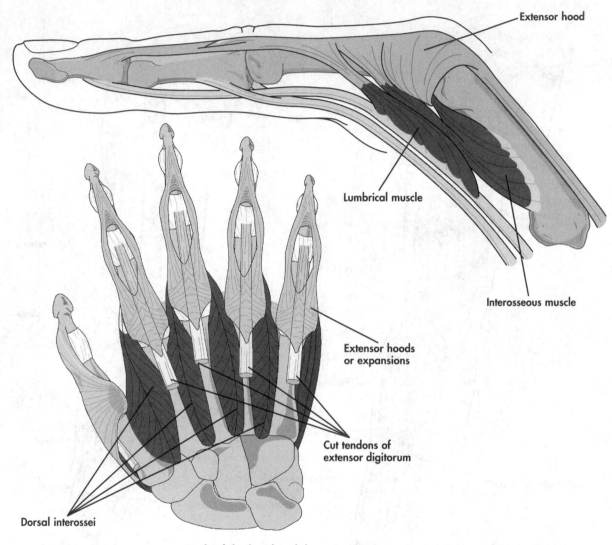

Back of the hand and the extensor expansions.

The tendons of the extensor digitorum cross the back of the hand and attach to both the middle and distal phalanges of the fingers. As the tendons cross the knuckles (MCP joints), they widen to form extensor hoods or expansions. Small muscles in the hand attach to these hoods.

Muscles of the Hand

The hand contains two groups of muscles:

♦ **Extrinsic hand muscles.**   These reach the hand via long tendons of muscles located in the forearm. The extrinsic muscles provide power for a firm grip on objects (for example, as when clutching a baseball bat).

♦ **Intrinsic hand muscles.**   Small muscles that originate in the hand. The intrinsic muscles are best suited for delicate, precise movements of the thumb and fingers (for example, as when threading fishing line through a hook or playing the piano).

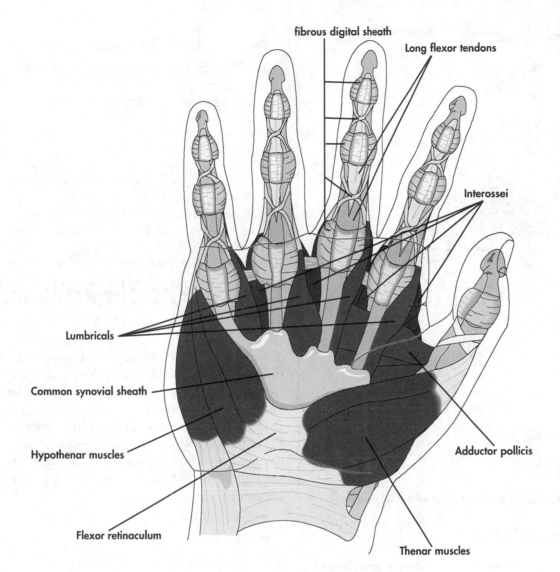

The intrinsic muscles of the hand.

At the base of the thumb and little finger near the wrist are prominences that represent the thenar and hypothenar muscle groups, respectively. The thenar muscles are the more important group because it contains a muscle that allows rotation of the thumb toward the other fingers (a movement called opposition) to produce a pinching action.

The lumbrical muscles are slender "worm-like" muscles that produce flexion at the MCP joints and extension at the IP joints of the fingers. These muscles are very important in modifying the opposing pulls of the long flexor and extensor tendons of the fingers to produce smooth, coordinated movements of the fingers.

Finally, the interossei muscles are responsible for spreading the fingers apart and for bringing them together. An easy way to remember which interossei do this is the following: The Palmar interossei ADduct the fingers ("PAD"), whereas the Dorsal interossei ABduct the fingers ("DAB").

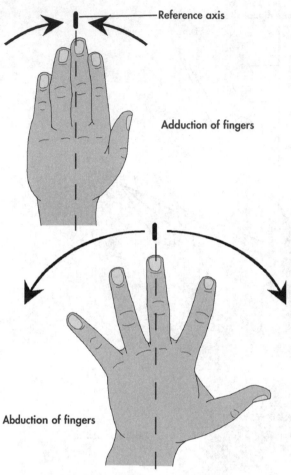

Abduction and adduction of the fingers.

Bodily Malfunctions

Repetitive hand activities, such as key-boarding or certain industrial work, may cause inflammation and swelling of the synovial sheath or tendons within the carpal tunnel. This puts pressure on the median nerve and produces tingling or numbness about the thumb, index, and mid-dle fingers. There may also be a loss of fine thumb movement, such as difficulty pinching or buttoning a shirt. This common overuse injury of the hand is known as carpal tunnel syndrome.

Dem's Da Pits: The Axilla and Its Contents

The nerve and blood supply to the upper limb originates in the axilla, or armpit. The axilla is located from side to side between the shoulder and trunk, and from front to back between the chest and back muscles. Through this space travel blood vessels, nerves, and lymphatic channels from the neck and thoracic cavity to the upper limb. The axilla contains the following components:

- The axillary artery
- The axillary vein
- The brachial plexus of nerves
- The axillary lymph nodes
- Lots of fat (as packing material)

This section summarizes the basic pattern of blood supply to the upper limb, as well as the nerve supply to this region.

Blood Supply to the Upper Limb

The principal source of blood to the upper limb is the *axillary artery*. It is a continuation of the subclavian artery, which passes out of the tho-rax. As the subclavian artery crosses the first rib, its name changes to axillary artery.

The Carpal Tunnel

The tendons of the flexor muscles to the fingers and thumb enter the palm of the hand through an enclosed space at the wrist, called the *car-pal tunnel*, before attaching to their respective bones. This fibrous-bony tunnel is formed by a tough sheet of connective tissue, the flexor retinaculum, which attaches to and arches across some of the carpal bones. The tendons are enclosed in a synovial sheath, which permits them to move freely within this confined space. The median nerve also passes through the carpal tunnel, between the flexor retinaculum and these flexor tendons. The ulnar nerve and artery, on the other hand, do not pass through the carpal tunnel.

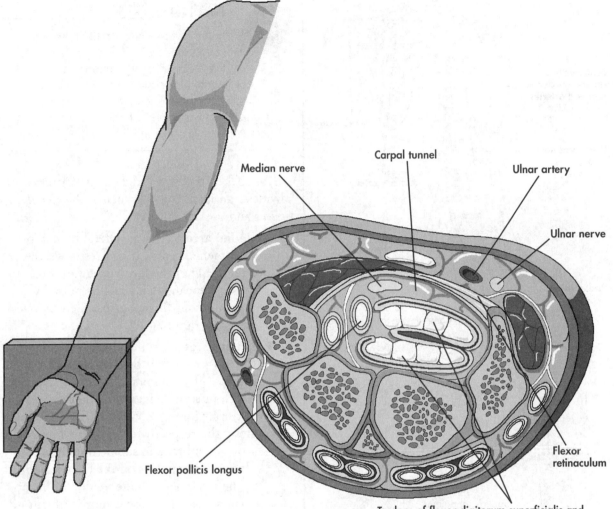

Median nerve

Carpal tunnel

Ulnar artery

Ulnar nerve

Flexor retinaculum

Flexor pollicis longus

Tendons of flexor digitorum superficialis and profundus enclosed in a common synovial sheath

The carpal tunnel.

Body Language

The primary artery to the upper limb is the **axillary artery.** It's a continuation of the subclavian artery and gives rise to several smaller arteries, such as the two circumflex humeral arteries.

Within the axilla, the axillary artery gives off several branches to muscles and structures of the shoulder and pectoral regions, including the anterior and posterior circumflex humeral arteries that encircle the surgical neck of the humerus. The posterior circumflex humeral artery is the larger of these two. It passes behind the surgical neck with the axillary nerve. These vessels supply the shoulder joint and adjacent muscles. At the lower border of the teres major muscle, the axillary artery changes its name to the *brachial artery* and enters the arm.

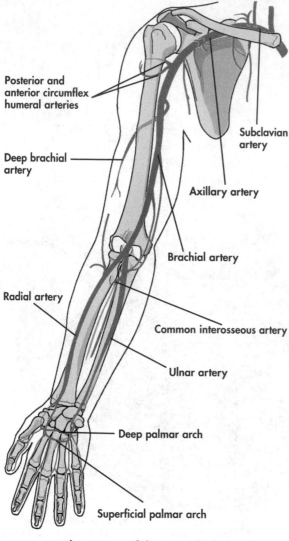

Posterior and
anterior circumflex
humeral arteries

Deep brachial
artery

Subclavian
artery

Axillary artery

Brachial artery

Radial artery

Common interosseous artery

Ulnar artery

Deep palmar arch

Superficial palmar arch

The arteries of the upper limb.

The brachial artery passes into the anterior
compartment of the arm on its medial side.
Early in its descent, it gives off the *deep brachial
artery*. This vessel passes into the posterior
compartment of the arm with the radial nerve.
Both of these follow within the radial groove
located at the midshaft of the humerus. The
artery supplies the triceps muscle and assists
in supplying blood to the elbow. The brachial
artery provides branches to the muscles in the
anterior compartment of the arm and to the
elbow.

Foot Notes

In the depression in front of the elbow,
called the *cubital fossa*, you can feel
the brachial artery pulse by placing
your fingers just medial to a thick, cordlike
structure that is the tendon of the biceps brachii. Blood pressure is measured by placing
a stethoscope over the artery here.

Just below the elbow, the brachial artery
divides into its terminal branches, the *radial
artery* and the *ulnar artery*:

◆ **Radial artery.**   Is the lateral branch of
the brachial artery. It courses down the
lateral side of the forearm toward the
wrist. You can readily feel its pulse by
pressing your finger against the skin overlying the radius.

◆ **Ulnar artery.**   Is the medial branch of
the brachial artery. Shortly after its origin,
it gives off a common *interosseous artery*,
which in turn divides into anterior and
posterior interosseous arteries that follow
along the respective sides of the interosseous membrane to supply muscles. The
ulnar artery appears on the medial side
of the wrist beneath the tendon of flexor
carpi ulnaris. Its pulse is more difficult to
feel.

The hand is supplied by branches of the
radial and ulnar arteries, which join to form a
pair of arterial arches: the superficial and deep
palmar arterial arches. Branches come off these
arches to supply the hand and fingers.

Venous blood from the upper limb returns
toward the heart by two sets of veins. One set,
called *superficial veins*, is located within the loose
connective tissue layer beneath the skin. These
are veins that are commonly visible through
the skin. A set of deep veins are located within
the upper limb and course with the arteries.

The superficial and deep veins are periodically connected by *perforating veins*. Both sets of veins contain valves that direct blood flow in the direction of the heart.

Foot Notes

Because they are close to the skin and can be made to stand out by applying local pressure, superficial veins are commonly used for drawing blood for analysis or for delivering fluids or drugs into the body. Veins on the back of the hand (the dorsal venous arch) or at the front of the elbow (the median cubital vein) are commonly used sites.

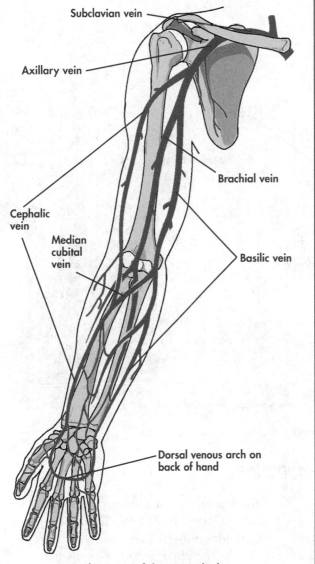

The veins of the upper limb.

The Nerve Supply to the Upper Limb

Spinal nerves from the C5, C6, C7, C8, and T1 levels of the spinal cord enter the axilla from the neck and form a network of nerves called the *brachial plexus*. From this plexus arise the nerves that supply the muscles and skin of the upper limb. The major branches of this plexus are the following:

◆ **The axillary nerve.** Supplies the deltoid and teres minor muscles. After leaving the brachial plexus, it passes behind the surgical neck of the humerus. This nerve may be injured during fractures of the upper humerus, resulting in weakness or paralysis of these muscles. In addition, the axillary nerve also supplies an area of skin along the upper lateral surface of the arm.

◆ **The radial nerve.** Leaves the brachial plexus and enters the posterior compartment of the arm where it supplies the triceps brachii muscle. It travels within the radial groove on the back side of the mid-shaft of the humerus (along with the deep brachial artery). The radial nerve divides in the forearm into a deep branch that is the motor supply to the muscles of the extensor forearm, and a superficial branch that is sensory to the skin on the back of the hand, particularly in the area between the first and second metacarpal bones (associated with the thumb and index finger, respectively).

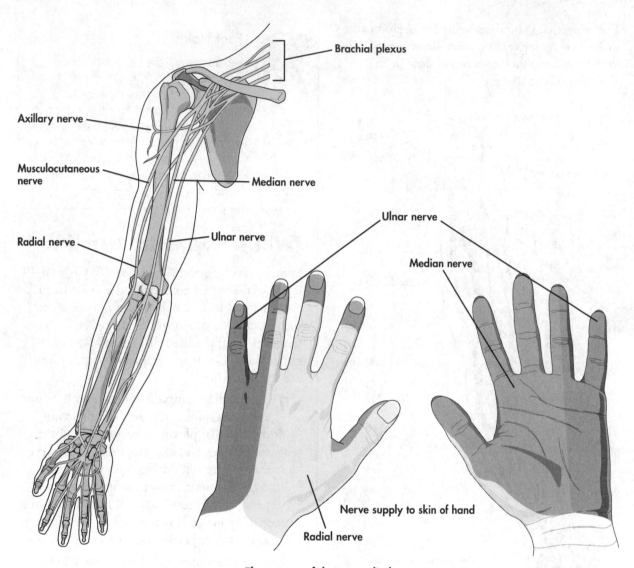

The nerves of the upper limb.

◆ **The musculocutaneous nerve.** The nerve of the anterior compartment of the arm. It supplies the coracobrachialis, brachialis, and biceps brachii muscles. It continues as a sensory nerve and supplies skin along the outer surface of the forearm.

◆ **The median nerve.** Courses with the brachial artery through the arm. It provides no branches in the arm. The median nerve enters the flexor compartment of the forearm, where it supplies all of these

muscles, except for the flexor carpi ulnaris and the lateral half of the flexor digitorum profundus muscle (these muscles are supplied by the ulnar nerve). The median nerve passes into the palm of the hand within the carpal tunnel before dividing into branches that supply the thenar muscles of the thumb and skin of the thumb, index, middle, and lateral half of the ring finger. The median nerve is a very important nerve of the hand because it supplies the intrinsic thumb muscles and much of the skin of the hand. In fact, it is described as being the "eyes of the hand" because it can enable a person to identify an object by feel.

♦ **The ulnar nerve.** Also courses through the arm without giving off any branches. In its descent, it passes behind the medial epicondyle of the humerus before it enters the flexor compartment of the forearm. The ulnar nerve supplies only the flexor carpi ulnaris and the medial half of the flexor digitorum profundus in the flexor compartment of the forearm. It descends down the medial aspect of the forearm and enters the palm of the hand outside of the carpal tunnel. It gives rise to branches in the hand that supply most of the intrinsic muscles of the hand, and supplies skin of the medial aspect of the palm, the little finger, and medial half of the ring finger. So its sensory supply to the hand is rather limited.

Foot Notes

While behind the medial epicondyle, the ulnar nerve is vulnerable to occasional bumps to the elbow. This classically produces pain that extends down the forearm to the little finger. This sensation is called "hitting the funny bone." The particular sensory experience has nothing to do with the bump to the bone; rather, it is a response by the nerve to this jolt.

The Least You Need to Know

♦ The upper limb consists of muscles, bones, and joints that function to place the hand in the optimal position to perform work.

♦ The upper limb is divided into the shoulder, arm, forearm, and hand regions.

♦ Muscles that flex the upper limb are generally located on the front side of the arm and forearm.

♦ Muscles that extend the upper limb are generally located on the back side of the arm and forearm.

♦ The nerve and blood supply of the upper limb originates in the axilla.

♦ The upper limb is not designed to be a weight-bearing structure. Efforts to break a fall with an outstretched hand often cause fractures or dislocations in any of the segments of the upper limb.

In This Chapter

- ◆ The relationship of the breast to the chest wall and its lymph drainage

- ◆ The organs within the thoracic cavity

- ◆ The bony and muscular parts of the thoracic wall

- ◆ The layers and components of the pleural and pericardial cavities

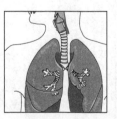
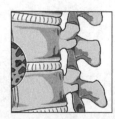
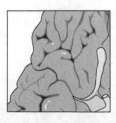

The Pectoral Region and Thorax: The Chest Area

Just as we place our most prized and valued possessions in safe places, like in a bank vault or security box, the body houses and protects its most vital organs (like the brain, the heart, and the lungs) within bony cavities.

The thorax is that part of the body between the neck and the abdomen. It consists of a bony cage, made up of 12 pairs of ribs, the sternum, and the thoracic vertebrae. A large and powerful muscle called the diaphragm attaches to the bottom edge of this bony cage and separates the thoracic cavity from the abdominopelvic cavity. The diaphragm is used in breathing. Between the ribs are layers of muscle that, along with the diaphragm, help change the size and dimensions of the rib cage. This bony-muscular wall contains the heart and lungs and serves two purposes: protection and respiration.

In this chapter, you will learn about the structures inside the thorax and about the breast and pectoral region that overlie the upper portion of the thoracic wall.

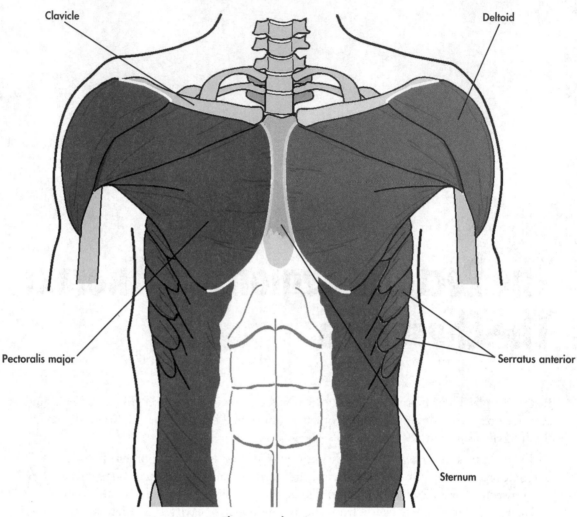

Clavicle

Deltoid

Pectoralis major

Serratus anterior

Sternum

The pectoral region.

On the Surface of Things: The Breast and Pectoral Region

The wall of the thorax, from the front and wrapping around to the back, is largely covered by muscles associated with the upper limb and with the back. These muscles are described in Chapters 15 and 16. The outer surface of the upper thorax is an area called the *pectoral region*. It includes the pectoralis major muscle, an upper limb muscle, and overlying it, the breast.

This area extends from the clavicle (collarbone) to about the sixth rib and from the breastbone to the armpit.

The breast is unremarkable in the male (and in the prepubertal female). It consists of the nipple, the pigmented area around the nipple called the areola, and some basic nonfunctional glandular tissue. These are located at the fourth intercostal space along a vertical line that passes through the midpoint of the clavicle (a.k.a. the mid-clavicular line). The adult female breast, on the other hand, is a prominent feature of the pectoral region and is clinically very important because of its vulnerability to disease, including cancer. Chapter 13 includes more detail on the anatomy and function of the female breast.

The female breast normally moves freely over the pectoralis major muscle. Several bands of connective tissue stretch from the deep portion of the breast to the skin and are called the *suspensory ligaments* of the breast. They provide support for the breast. Fat fills the spaces between the fibrous and gland tissue and gives the breast its rounded shape.

Bodily Malfunctions _____

Occasionally, the suspensory ligaments of the breast become involved in invasive cancer. When this happens, the connective tissue bands shorten, producing a dimple or pucker on the skin surface. Such a pucker is one of the warning signs of cancer that women are told to look for.

The Quadrants of the Breast

Clinically, it is common to divide the breast into quadrants by right-angle lines that intersect at the nipple. The four quadrants are the following:

◆ Upper inner

◆ Upper outer

◆ Lower inner

◆ Lower outer

These quadrants provide a helpful map for reporting locations of lumps, pain, or infections in the breast. The upper outer quadrant contains an extension of breast tissue, called the axillary tail, which wraps around the lower border of the pectoralis major muscle and joins the contents of the axilla.

Lymphatic Drainage of the Breast

In Chapter 8 you learned that lymph is a tissue fluid that is reabsorbed by lymph capillaries and transported to a nearby lymph node. There it is cleansed of bacteria and foreign material, including cancer cells. Because cancer cells commonly leave their original tumor sites and spread (metastasize) to distant tissues by way of the blood and lymph streams, it is important for doctors to know the pathway cancer cells take from a particular organ.

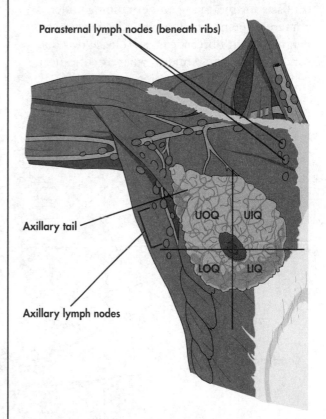

The lymphatic drainage of the breast.

Lymph from the outer quadrants of the breast commonly flows toward lymph nodes located within the axilla, the axillary lymph nodes. Lymph from the inner quadrants generally flows toward the *parasternal lymph nodes* located inside the chest cavity. These nodes are not easy to access because they lie beneath the ribs where they attach to the sternum. Occasionally, lymph may flow to lymphatic channels in the opposite breast or to lymphatic channels on the abdominal wall. About 75 percent of the lymph from the breast drains to the pectoral group of axillary lymph nodes, located along the outer border of the pectoralis minor muscle beneath the pectoralis major muscle. The remaining 25 percent of the lymph passes into the parasternal nodes. This is the clinical basis for biopsying and examining under a microscope nearby lymph nodes in patients diagnosed with breast cancer. This allows doctors to determine if and to what extent cancer cells have spread from a breast tumor.

The Thoracic Wall

The skeleton of the thoracic wall consists of 12 thoracic vertebrae in back, the sternum in front, and 12 pairs of ribs in between. At the end of each rib is a segment of hyaline cartilage, called the *costal cartilage*. All 12 pairs of ribs connect to the thoracic vertebrae. However, not all of the ribs attach to the sternum at the other end. The first seven ribs are called "true ribs" because they attach to the sternum. The rest are called "false ribs." Ribs 8–10 attach to the costal cartilage of the rib above the other. Ribs 11 and 12 are called "floating" ribs. Their free ends are embedded in muscle of the body wall. Where the costal cartilages of ribs 8, 9, and 10 join is called the *costal margins* of the rib cage. The union of the right and left costal margins to the lower end of the sternum forms the *costal arch*.

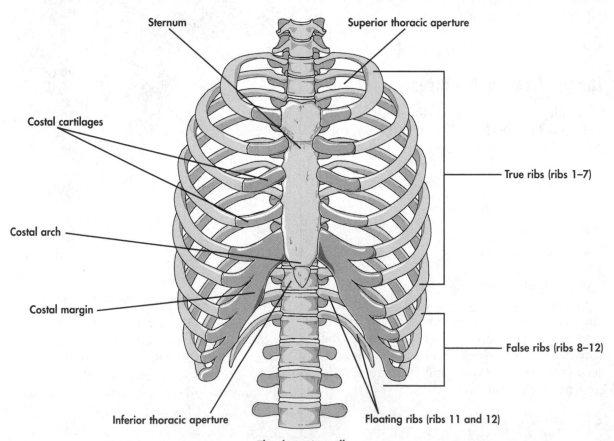

The thoracic wall.

Openings of the Thoracic Cavity

There are two openings at opposite ends of the thoracic cavity. Toward the neck is the *superior thoracic aperture* (or opening). Into the sides of this opening project the apex of each lung and the subclavian arteries and veins, which leave the thoracic cavity and course to the neck and upper limb. In the midline of this opening are structures that enter and leave the neck, such as the trachea, the esophagus, the common carotid arteries, the internal jugular veins, nerves, and the thoracic duct.

The *inferior thoracic aperture* is closed by a fibromuscular structure called the *respiratory diaphragm*. The diaphragm's central portion is made of the same material as tendons and attaches to the pericardial sac that contains the heart. Its edges are muscular and attach to the sternum, the twelfth thoracic vertebra, and along the costal arch. When it's relaxed (when the lungs are empty), the diaphragm appears dome-shaped. The dome is slightly higher on the right side, where it rises up because of the liver beneath it. During contraction, the diaphragm descends and flattens out to increase the volume of the thoracic cavity to accommodate more air entering the lungs. The nerve supply to the diaphragm is the *phrenic nerve*.

The diaphragm has three openings: for the inferior vena cava, the esophagus, and the aorta. These structures enter and leave the abdomino-pelvic cavity.

The Ribs and the Intercostal Spaces

The space between each pair of ribs, the *intercostal space*, is filled with three more or less continuous layers of intercostal muscles. The intercostal muscles are involved in breathing and are discussed later in this chapter. Also associated with these spaces are intercostal nerves and blood vessels.

Body Language _____

Costal is a Latin word meaning "of or near the ribs."

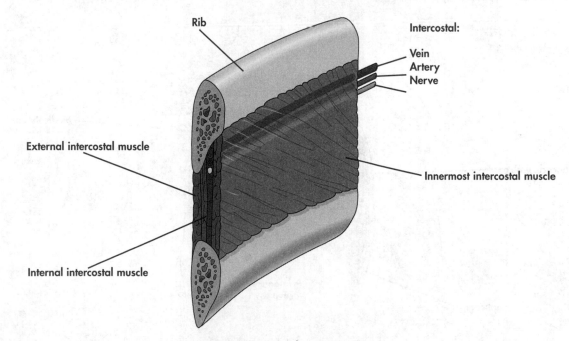

An intercostal space.

The nerve supply to these muscles is by way of (guess what?) *intercostal nerves* (don't you wish all of anatomy was this easy?). Each intercostal nerve is actually a ventral primary ramus (see Chapter 6) that leaves its respective intervertebral foramen and enters an intercostal space containing muscles. These nerves pass all the way to the front of the body. They supply motor fibers to the muscles within their space and sensory fibers to the overlying skin. There are 11 pairs of intercostal nerves (T1–T11). The ventral primary ramus of T12 courses below the last rib and is called the *subcostal nerve*. Intercostal and subcostal nerves T7–T12 course onto the abdominal wall once they leave their intercostal spaces. They also supply muscle and skin in that region.

The pattern and course of blood supply to the intercostal muscles and tissue spaces generally follows that of the nerves. The principal blood supply is provided by posterior intercostal arteries. These are primarily branches of the thoracic aorta. As these vessels course forward, they connect with anterior intercostal arteries that arise from internal thoracic arteries. These,

in turn, are branches of the subclavian arteries. They pass downward beneath the ribs near their attachment to the sternum. At the end of the sternum, they divide into the *musculophrenic* and *superior epigastric arteries*.

Venous return from these intercostal spaces is by way of corresponding anterior and posterior intercostal veins. The posterior intercostal veins drain into the *azygous* system of veins located along the back of the thoracic cavity, which return venous blood from much of the thoracic wall to the superior vena cava.

Whatcha Got Under the Hood?: The Thoracic Organs

Beneath the anterior thoracic wall is a broad space called the *thoracic cavity*. This cavity is divided into three compartments. First is a central cavity called the *mediastinum*, which contains the heart, the trachea, the esophagus, and other structures. There are also two pulmonary cavities, which contain the lungs and their pleural membranes.

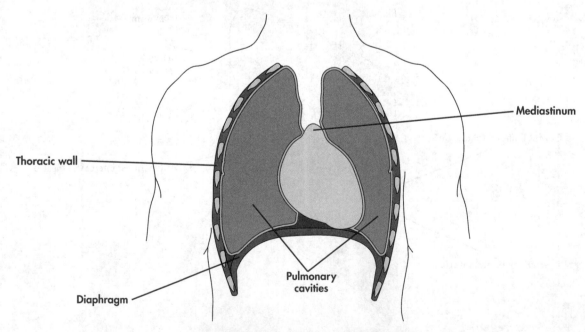

The compartments of the thoracic cavity.

The Heart

The mediastinum is the central region of the thoracic cavity, located between the right and left pulmonary cavities. Its principal structure is the heart, contained within its pericardial sac.

Visceral layer of serous pericardium (epicardium)

Parietal layer of serous pericardium (applied to fibrous pericardium)

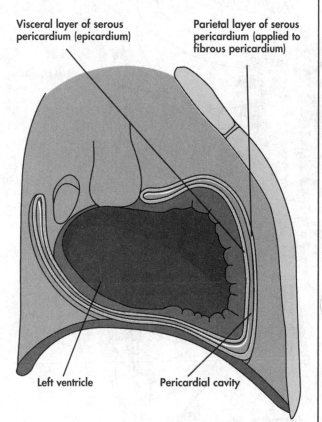

Left ventricle

Pericardial cavity

The pericardium of the heart.

The pericardial sac consists of two layers:

◆ **Fibrous pericardium.** The outer tough fibrous layer. It attaches above to the outer surfaces of the great vessels of the heart and below to the central tendinous portion of the diaphragm.

◆ **Serous pericardium.** Composed of a layer of mesothelial cells. It consists of two portions: the parietal layer of serous pericardium, which lines the internal surface of the fibrous pericardium, and the visceral layer of serous pericardium, which covers the external surface of the heart. This layer is also called the *epicardium* of the heart.

The two layers of serous pericardium enclose the *pericardial cavity*. The pericardial space contains a small amount of serous fluid to allow the heart to beat without friction.

Bodily Malfunctions

Cardiac tamponade is a potentially life-threatening condition involving a sudden accumulation of blood or fluid within the pericardial cavity, as for example, following a stab wound or bleeding from a ruptured coronary artery. The tough but inelastic fibrous pericardium causes an increase in pressure within this cavity that compresses the heart, preventing it from pumping effectively.

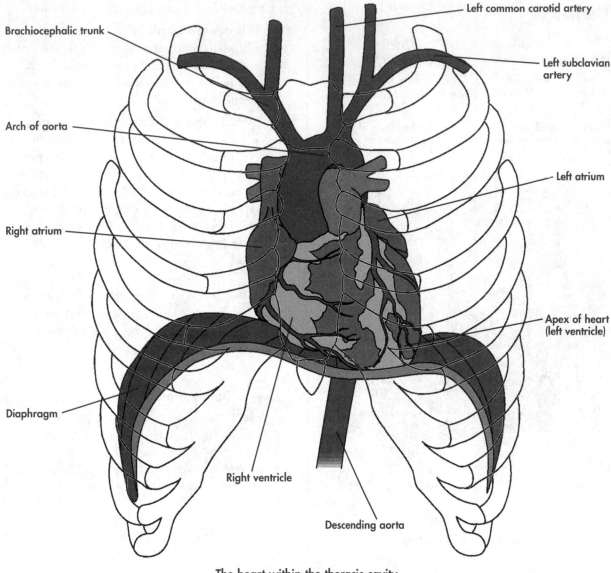

Brachiocephalic trunk

Arch of aorta

Right atrium

Diaphragm

Left common carotid artery

Left subclavian artery

Left atrium

Apex of heart (left ventricle)

Right ventricle

Descending aorta

The heart within the thoracic cavity.

Refer to Chapter 7 for the details of the external and internal anatomy of the heart. The position and location of the heart and its various chambers, if you had "x-ray vision" to see through the skin of the anterior chest wall, are as follows:

◆ **Right margin of the heart.** Formed mainly by the right atrium, this area extends along the right side of the sternum between the third and sixth ribs.

◆ **Lower margin of the heart.** Formed mainly by the right ventricle. It extends from the right sixth rib near the sternum to the apex of the heart, located in the left fifth intercostal space near the mid-clavicular line.

◆ **Left margin of the heart.** Formed mainly by the left ventricle. It extends from the fifth intercostal space near the mid-clavicular line to the left border of the sternum near the second intercostal space.

◆ **Superior border of the heart.** Consists of an area formed by several of the great vessels. The superior vena cava is formed when the right and left brachiocephalic veins join. These veins begin behind the right and left sternoclavicular joints by union of an internal jugular vein (carrying blood from the brain, head, and neck) with a subclavian vein (returning blood from the upper limb and shoulder).

Shortly after leaving the left ventricle of the heart, the aorta arches toward the left as the aortic arch before coursing downward as the descending or thoracic aorta. Coming off the arch of the aorta are three branches:

◆ **The brachiocephalic trunk.** The largest and the first branch. It divides behind the right sternoclavicular joint into the right common carotid artery and the right subclavian artery.

◆ **The left common carotid artery.** Both right and left common carotid arteries pass upward into the neck.

◆ **The left subclavian artery.** The right and left subclavian arteries also enter the neck through the superior thoracic opening and give rise to branches in the neck. The subclavian arteries cross the first rib to become the axillary arteries that supply blood to the upper limb.

The Lungs and Pleura

The membranes that surround the lungs are called the *pleura*. The cavities include two layers of pleura. Lining the surfaces of each pulmonary cavity is a thin layer of *parietal pleura*. The parietal pleurae have four parts:

Body Language _____

Pleura are the membranes that cover the lungs and line the pulmonary cavity. The word is derived from a Greek word meaning "rib or side."

◆ **Costal pleura.** The part that lines the internal surface of the thoracic wall.

◆ **Diaphragmatic pleura.** The part that covers the upper surface of the diaphragm.

◆ **Mediastinal pleura.** The part that covers the fibrous pericardium overlying the heart.

◆ **Cervical pleura or cupola.** The part that extends up a short distance into the neck and covers the upper extent of the pulmonary cavity.

The *visceral pleura* is stuck to the outside surface of each lung and encloses them. At the hilum of the lung, the visceral pleura is continuous with the mediastinal pleura. The pleural cavity represents the potential space between the parietal and visceral layers of pleura. Normally, this space contains a small amount of serous fluid. This fluid is produced by mesothelial cells that line the pleural surfaces. It lubricates the pleural surfaces and allows smooth, pain-free movement between them during breathing.

Foot Notes _____

The concept of "potential space" is difficult to grasp. Imagine that you are pushing your fist (representing the lung) into a balloon (representing the pleural cavity). There would be no space between the lung and the pleural cavity. But if something happened to the lung and it bled into the cavity, the potential space would become a real space, filled with blood.

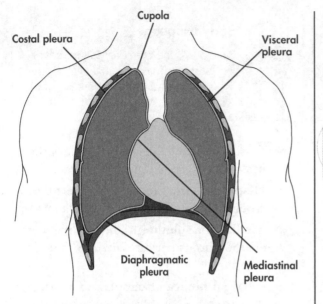

Costal pleura

Cupola

Visceral pleura

Diaphragmatic pleura

Mediastinal pleura

The pleural membranes of the lungs.

Bodily Malfunctions

Pleurisy, or pleuritis, is an inflammation of the pleura. There are many causes, including viral infection or pneumonia. The inflammation causes the pleural membranes to become roughened and they rub against one another, making breathing very painful.

The lungs are divided by fissures (grooves) into unequal parts called lobes. The right lung has three lobes: an upper, a middle, and a lower lobe. The left lung has two lobes, an upper and a lower lobe. The surfaces of each lung lie next to ribs, the diaphragm, or the mediastinum, or project into the lower neck. The *hilum*, or root of the lung, is located at its mediastinal surface. Here, structures enter and leave the lung, such as the primary bronchi, pulmonary arteries and pulmonary veins, lymph nodes and lymphatic vessels, and autonomic nerves.

The trachea divides in the upper part of the thoracic cavity into right and left primary bronchi. The right primary bronchus is wider, shorter, and more vertical than the left primary bronchus.

Bodily Malfunctions

A foreign body that is inhaled and sucked down the windpipe (trachea) is more likely to enter the right bronchus than the left because it is more in line with the direction of the trachea and is larger in diameter than the left bronchus.

Each primary bronchus divides into a secondary or lobar bronchus that enters the individual lobes of the lung. The right lung has three secondary bronchi; the left lung has two. Within each lobe, the secondary bronchi divide into tertiary or segmental bronchi that provide air to specific segments of the lobe. These are called *bronchopulmonary segments*. There are generally 10 bronchopulmonary segments in each lung. These bronchi subsequently branch into smaller and smaller airways until you get down to the alveoli, where gaseous exchange happens. The details of the lower respiratory tree are discussed in Chapter 9.

The pulmonary arteries, which arise from the pulmonary trunk of the heart, carry deoxygenated blood to the lungs. Each pulmonary artery enters a lung hilum and divides into lobar and segmental arteries that accompany their respective bronchi to the alveoli for oxygenation. Pulmonary veins carry oxygenated blood from each lung to the left atrium of the heart.

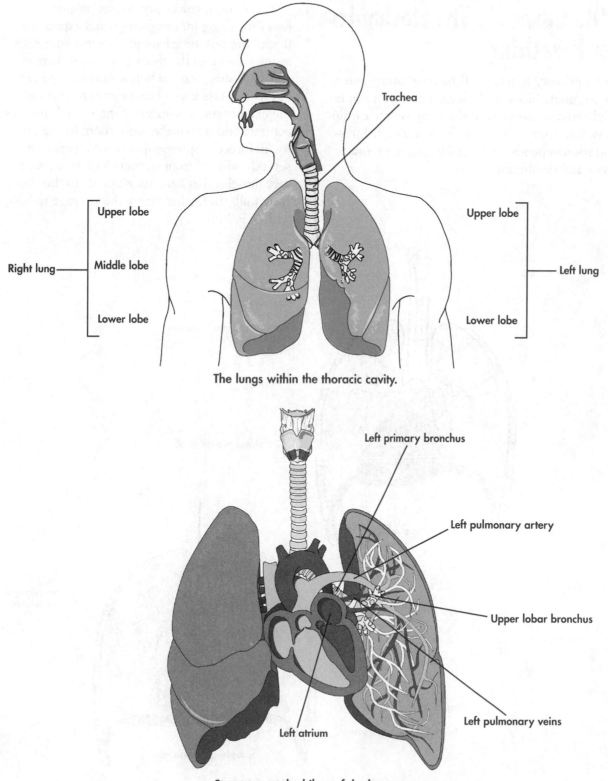

The lungs within the thoracic cavity.

Structures at the hilum of the lung.

"He Gasped": The Mechanics of Breathing

The primary function of the respiratory system is to supply blood with oxygen so that it can be delivered to the cells of the body via the cardiovascular system. The ability to accomplish respiration depends on a flexible, mobile thoracic wall and diaphragm.

Breathing involves two phases: inspiration (breathing in, or inhaling) and expiration (breathing out, or exhaling). Inspiration occurs as the volume of the thoracic cavity is increased and air rushes into the lungs. About 65 percent of this increase is produced by contraction of the diaphragm. The external intercostal muscles contract and help make more room in the cavity. In forced inspiration, as when a person is severely winded from strenuous exercise, accessory muscles that have attachments to the thoracic wall can further lift up the rib cage to help with breathing.

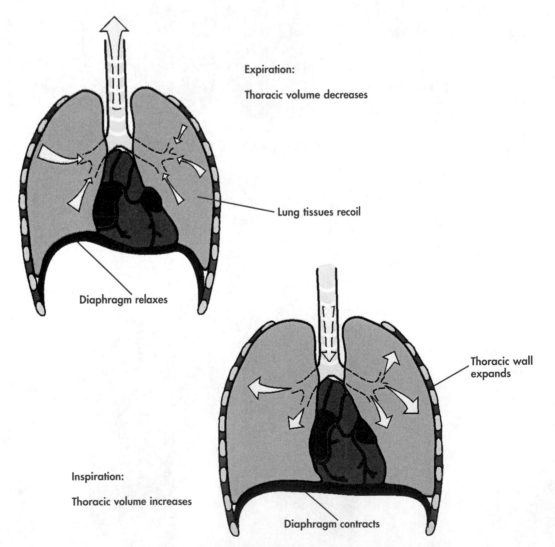

Expiration:

Thoracic volume decreases

Lung tissues recoil

Diaphragm relaxes

Inspiration:

Thoracic volume increases

Thoracic wall expands

Diaphragm contracts

The mechanics of breathing.

Expiration, on the other hand, is when air flows out of the lungs. This is largely a passive event that involves relaxation of the diaphragm and elastic recoil of the lungs themselves. Forced expiration, as when one tries to blow up a balloon or a cake-full of candles, involves contraction of the internal intercostal muscles and the abdominal wall muscles (described in Chapter 18).

Foot Notes

The sole motor nerve supply to the diaphragm are the phrenic nerves. They arise from cervical spinal nerves C3, C4, and C5. Fractures of the neck that severely injure the spinal cord above the origin of these nerves mean that the diaphragm will be paralyzed (as happened to actor/activist Christopher Reeve). Such individuals require external ventilators to "breathe" for them.

The Least You Need to Know

◆ The adult female breast contains the mammary glands within the superficial fascia and moves freely over the pectoralis major muscle.

◆ Most of the lymph from the breast drains into the axillary lymph nodes.

◆ The thoracic cavity consists of a mobile bony-muscular wall that is open at the neck, for structures entering and leaving the thorax, and closed below by the respiratory diaphragm, which separates this cavity from the abdominopelvic cavity.

◆ The lungs and heart lie within the thoracic cavity.

In This Chapter

- ◆ The front wall of the abdomen
- ◆ The gastrointestinal tract and its organs
- ◆ Accessory organs of the digestive system
- ◆ The rear wall of the abdomen

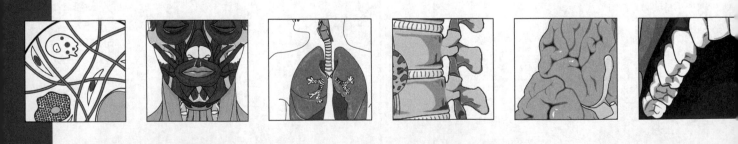

18

The Abdomen: The Belly of the Beast

The abdomen, also known as the belly, is the region of the body between the thorax and the pelvis. It is a cavity that contains several abdominal organs, such as the stomach, the small and large intestines, the liver, the spleen, the pancreas, and the kidneys. This cavity opens directly into the pelvic cavity. It is the major part of the abdominopelvic cavity that is bounded above by the respiratory diaphragm and below by the pelvic diaphragm.

Clinically, the abdomen is a very important region of the body. It contains many of the body's most vital organs for supplying the body with energy. Also it is an area that frequently experiences disease and trauma. This chapter should help you understand the basic organization and relationship of structures in this large cavity of the body.

Scaling the Wall: The Anterior (Front) Abdominal Wall

The walls of the abdominal cavity are formed by bone and skeletal muscle, which are covered on the outside by skin and connective tissue (called *superficial fascia*) and on the inside by a sac lining (called *peritoneum*). The bony boundaries are as follows:

- ◆ Above, by the sternum, costal margins, ribs 11 and 12, and twelfth thoracic vertebra
- ◆ In back, by the five lumbar vertebrae
- ◆ Below, by the *pelvic brim*, which is the bony boundary at the circular opening between the abdominal and pelvic cavities. (It is made up of portions of the hip bones and the sacrum.)

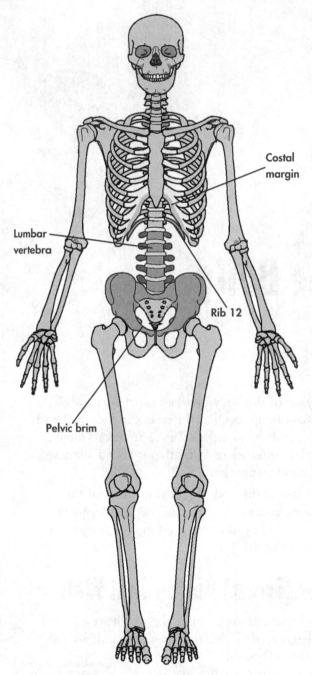

Bony boundaries of the abdominal cavity.

Because the front of the pelvis is slanted or tilted downward, the plane of the opening between the abdominal cavity and the pelvic cavity is about 90 degrees. As a result, the front wall of the abdomen is longer than the back wall.

Planes and Quadrants: "It Hurts Here!"

The anterior abdominal wall can be divided into quadrants by a vertical line that passes down the middle of the sternum and a horizontal line that intersects it at the bellybutton. These are the right and left upper quadrants and the right and left lower quadrants. This system is not so that games can be played on your tummy if you're bored; rather, such surface mapping helps both patient and doctor communicate about locations of pain, organs, or where surgical incisions will be made.

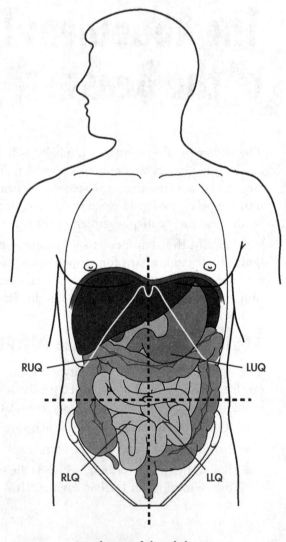

Quadrants of the abdomen.

Muscles of the Anterior Abdominal Wall

Remember in the previous chapter when we talked about the three layers of intercostal muscles that filled each space between the ribs? Well, there are three similar layers of muscles that form most of the anterior and lateral walls of the abdomen. These are the flank muscles. These muscles are very large and cover great areas of the body wall. The flank muscles have broad, flat tendons, called *aponeuroses*, which join together at the midline of the anterior abdominal wall. This tough connective tissue gives strength to the wall and protects the abdominal organs. A fourth layer of muscles, the rectus abdominis, extends up and down in the middle of the anterior abdominal wall. It is enclosed by the aponeuroses of the flank muscles.

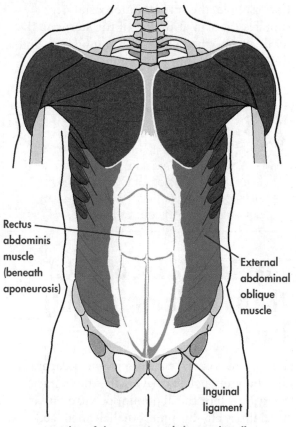

Muscles of the anterior abdominal wall.

Body Language

Aponeuroses are tendonlike sheets of connective tissue that attach muscles to bone or to other aponeuroses.

The anterior abdominal muscles are as follows:

- **External abdominal oblique.**   The outermost layer. It is the lower edge of its aponeurosis forms the inguinal ligament.
- **Internal abdominal oblique.**   The middle layer.
- **Transversus abdominis.**   The innermost muscle layer.
- **Rectus abdominis.**   Rectus refers to "straight." This pair of muscles extends from ribs five to seven above to the pubic bone below. Each pair is divided into four separate muscle bellies connected by tendinous insertions. Extensive development of these, through abdominal strength exercises, produce the "six-pack" abs!

The anterior abdominal wall muscles do the following:

- Provide strong, flexible support of the anterior abdominal wall
- Protect abdominal organs from injury
- Help flex and rotate the trunk
- Increase (when contracted) pressure within the abdominal cavity during emptying of urine from the bladder, feces from the rectum, and a baby from the uterus!

The nerve supply to the abdominal wall, including sensory supply to the skin and parietal peritoneum and motor supply to the abdominal muscles, is via the intercostal nerves (thoracoabdominal nerves). These nerves leave their intercostal spaces at the costal margins of the rib cage and continue into the abdominal wall.

For example, the seventh intercostal nerve supplies the skin at the tip of the sternum, the tenth around the bellybutton (umbilicus), and the twelfth just above the pubic bone.

Bodily Malfunctions

The lower anterior abdominal wall has holes in it called the deep and superficial inguinal rings. The superficial ring is a potential weak spot in the anterior abdominal wall where intestinal contents may protrude, or herniate.

That Takes a Lot of Guts!: The Gastrointestinal (GI) Tract

The *gastrointestinal tract* is a major occupant of the abdominopelvic cavity. Like the thoracic cavity, much of the abdominopelvic cavity is lined internally by a mesothelial-lined sac, called the *peritoneal sac*. Also just as in the thoracic cavity, this lining has two layers: the parietal layer and the *visceral layer*.

Body Language

The **gastrointestinal (or GI) tract** includes the esophagus, stomach, small intestine, and large intestine.

Parietal means "related to the wall of a cavity"; the parietal layer of peritoneum lines the inner surfaces of the abdominal walls (the underside of the diaphragm, the posterior abdominal wall and anterior abdominal wall). The **visceral** layer covers the surfaces of the organs.

Within the peritoneal sac is a small amount of serous fluid that allows the abdominal organs to move about smoothly within the cavity. In addition to its ability to secrete serous fluid, the peritoneum can absorb fluid. The peritoneal cavity is a completely closed space in the male, but is open in the female at the uterine tubes.

Bodily Malfunctions

Inflammation of the peritoneal lining is called peritonitis. It is caused by bacteria, bile, or chemical irritants entering the peritoneal sac from a tear in an infected or diseased abdominal organ, such as from a burst appendix, a perforated ulcer, or contamination during surgery. Peritonitis can be a severe, life-threatening, and painful condition that requires prompt medical care.

The following sections detail the major parts of the GI tract contained within the abdominopelvic cavity.

Esophagus: Food Tube

In Chapter 10, you learned that the esophagus is part of the digestive system. It is a muscular tube that starts at the throat. It passes through the superior thoracic aperture behind the trachea into the thorax, where it descends in front of the thoracic spine. Following its passage through an opening in the diaphragm, it immediately connects to the cardia portion of the stomach. The end of the esophagus contains the lower esophageal sphincter that prevents air from entering and filling the stomach as well as preventing stomach contents from refluxing into the esophagus.

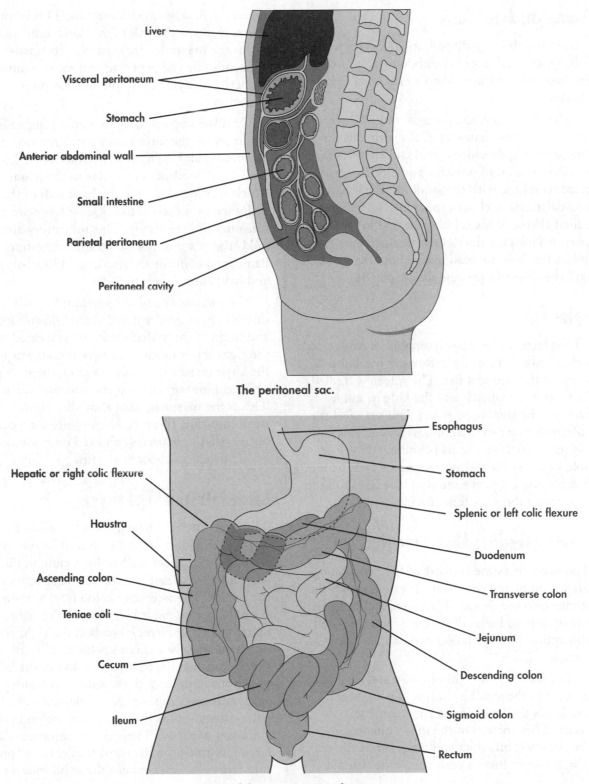

The peritoneal sac.

Liver

Visceral peritoneum

Stomach

Anterior abdominal wall

Small intestine

Parietal peritoneum

Peritoneal cavity

Parts of the gastrointestinal tract.

Esophagus

Hepatic or right colic flexure

Stomach

Splenic or left colic flexure

Haustra

Duodenum

Ascending colon

Transverse colon

Teniae coli

Jejunum

Cecum

Descending colon

Ileum

Sigmoid colon

Rectum

Stomach: Acid Wash

The stomach is a J-shaped organ located in the left upper quadrant of the abdomen. It varies in shape and size according to a person's food intake.

The stomach is the portion of the GI tract with the largest diameter. Cells within its lining secrete hydrochloric acid and enzymes that begin digestion of proteins and fats. Its gastric juices combine with the swallowed food and muscular action churns these into a semifluid, called chyme. This is later squirted in small portions into the duodenum through the pyloric sphincter. You can read more about the stomach and the digestive process in Chapter 10.

Spleen

The spleen is a vascular lymphoid organ located behind ribs 9–11 in the left upper quadrant. It is about the size of a fist. The spleen is cradled between the stomach and the kidney, and it rests on the transverse colon. You can't feel the spleen through the skin unless it is at least double its normal size. In its position, it is vulnerable to injury following fracture of a rib, where it can cause extensive internal bleeding. You can read more about its function in Chapter 8.

Small Intestine: Absorbing Nutrients

The small intestine consists of the duodenum, the jejunum, and the ileum. It extends from the stomach to the ileocecal junction, a distance of about 20 feet. Here, digestion of nutrients and absorption into the blood and lymph systems is completed.

The duodenum is the shortest and most fixed portion of the small intestine. It is C-shaped and holds the head of the pancreas in its concavity. This portion of the duodenum receives the common bile duct and main pancreatic duct, which fuse to form a common, wide channel, called the *hepatopancreatic ampulla*. This ampulla passes through the wall of the duodenum and forms the major duodenal papilla. Its opening sends bile and pancreatic juices, containing numerous enzymes, into the duodenum for digestion.

Circular mucosal folds containing finger-like villi increase the surface area for digestion and absorption and are most numerous in the jejunum. The duodenum connects to the jejunum, which is suspended from the back wall of the abdomen by an apron-like double layer of peritoneum called *mesentery*. This mesentery also holds the next portion of the small intestine, the ileum, and contains the nerve and blood supply and lymphatic vessels to them.

The mesentery and the jejunum begin in the left upper quadrant and extend downward and to the right so that the ileum is located in the right lower quadrant, where it connects to the large intestine. At this junction, the ileum projects into the cecum at the ileocecal valve. This is the narrowest area along the entire small intestine. It can be lodging site for foreign objects that have been swallowed or gallstones. Either produces abdominal pain.

Large Intestine: The Dryer

The large intestine extends about five feet from the ileocecal junction to the anus. It begins as a blind-ended pouch, called the cecum, in the right lower quadrant. The appendix attaches to the cecum. The ascending colon travels upward from the cecum toward the liver in the right upper quadrant. Here, it bends at the right colic or hepatic flexure and crosses the midline of the abdomen as the transverse colon to the left upper quadrant beneath the spleen. It bends again at the left colic or splenic flexure and passes downward as the descending colon to the left lower quadrant. These three segments of the large intestine form the three borders of a "picture frame," which contains the small intestine.

The sigmoid colon is an S-shaped portion of the large intestine that leaves the left lower quadrant and passes over the pelvic brim into the pelvic cavity. Here it straightens out as the rectum. The rectum ends at the anal canal. (See Chapter 19 for more on the rectum.)

The mucosal lining of the large intestine specializes in absorbing water and salts and producing mucus for lubricating the intestinal lining to help the fecal mass pass toward the rectum.

Three distinguishing features characterize the large intestine:

◆ **Teniae coli.** These are three narrow bands of longitudinal smooth muscle located on the outer wall of the large intestine from the cecum to the rectum. This portion of the large intestine does not have a complete outer layer of longitudinal smooth muscle.

◆ **Haustra or sacculations.** These are accordion-like segments of the large intestine.

◆ **Epiploic appendages.** These are fat-filled structures that are enclosed in peritoneum. They hang off the large intestine along the edges of the teniae coli.

Bodily Malfunctions

Diverticuli are small herniations of the lining of the large intestine, usually in the sigmoid part. They occur along the edges of the teniae coli where the intestinal wall is weakest. The presence of these sacs, or diverticulosis, is an acquired disorder in the Western world due to low fiber in the diet. If the sacs become infected, it's called diverticulitis. This painful condition is treated with antibiotics, and patients are put on a special diet.

Just There to Help: Accessory Organs of the Digestive System

In addition to the GI tract, the abdominopelvic cavity is also home to the *accessory organs* that assist with digestion. The following sections give details on each of these organs.

Body Language

The **accessory organs** of the digestive system include the liver, the gallbladder, and the pancreas. They help digest and metabolize the nutrients from the digestive tract.

Liver: Metabolism and Detox

The liver is the largest visceral organ in the body, weighing about 1.5 kg. It is located principally in the right upper quadrant beneath the diaphragm, which it causes to dome higher on the right side. It narrows and extends across the midline to enter the left upper quadrant. It is largely protected by the rib cage. Its lower border does not usually extend beyond the right costal margin. This is why needle biopsies of the liver for microscopic examination of disease are performed through lower intercostal spaces (eighth through tenth) on the right lateral side of the body.

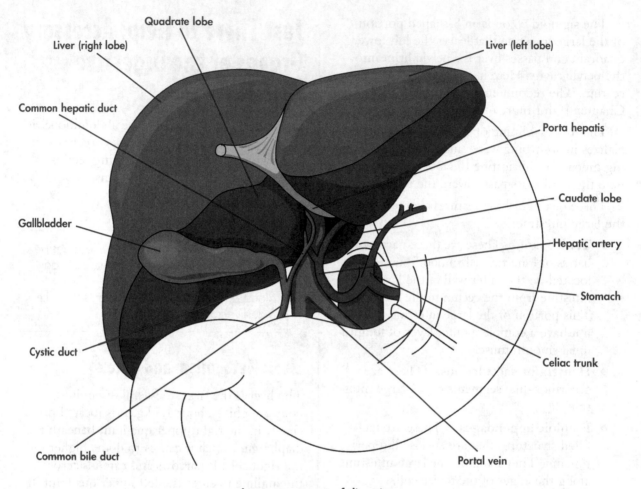

Accessory organs of digestion.

The liver is divided into four anatomic lobes: right lobe, left lobe, quadrate lobe, and caudate lobe. The right lobe is the largest of the lobes.

The porta hepatis is the site of entrance and exit of the bile ducts, the hepatic artery, and the portal vein. It is located on the undersurface of the liver. The gallbladder is attached here as well. The liver receives a rich blood supply from two sources: the portal vein, carrying nutrient-rich blood from the GI tract, and the hepatic artery, carrying oxygen-rich blood from the aorta, via its celiac trunk. In this strategic position, it metabolizes and detoxifies products of digestion. It also is an important producer of blood proteins and bile.

Gallbladder: The Storage Tank

The gallbladder is part of the biliary system, which consists of the liver, the gallbladder, and their ducts. Bile is produced in the liver, but the bile is stored and concentrated in the gallbladder until it's released during digestion. Bile leaves the right and left lobes of the liver through right and left hepatic ducts. A short distance thereafter they join as the common hepatic duct. The common hepatic duct joins the cystic duct from the gallbladder to form the common bile duct. This channel passes behind the first part of the duodenum and often passes through the head of the pancreas, where it unites with the main pancreatic duct. Together they enter the wall of the second part of the duodenum and open at the major duodenal papilla.

A portion of the gallbladder, called the fundus, projects from the lower border of the liver and intersects the outer edge of the rectus abdominis muscle as it crosses the right costal margin. An inflamed, tender gallbladder may be felt at this intersection.

It is not uncommon for the biliary system to become blocked by gallstones. A typical site for blockage is where the common bile duct enters the duodenum. Swelling of the duct causes severe pain until the stone is passed or surgically removed. Cancer of the head of the pancreas may block flow of bile into the duodenum, causing the yellowish bile pigments to be absorbed into the bloodstream. This produces the condition called jaundice, which discolors the skin and the "whites" of the eyes.

Pancreas: The Juice Bar

The pancreas is a glandular organ that has both exocrine and endocrine functions. It is located behind the peritoneum against the back wall of the abdomen. The transverse colon and stomach lie in front of it.

The exocrine portion of the gland produces several kinds of enzymes involved in the breakdown of starches, fats, and proteins. These "external" secretions are transported to the duodenum via the main pancreatic duct. The pancreas' endocrine (or hormonal) portion is contained within clusters of cells called the *islets of Langerhans*. These cells produce "internal" secretions into the bloodstream. The principal hormones produced are insulin and glucagon, which is used in regulating blood sugar levels. Inadequate production or production of abnormal insulin causes blood sugar levels to elevate beyond their normal range. This condition is known as diabetes mellitus.

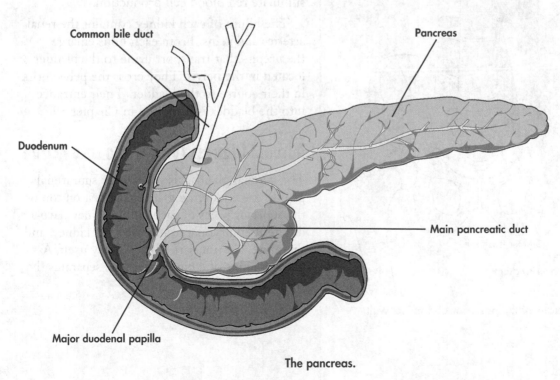

The pancreas.

Up Against a Wall: Organs of the Posterior Abdominal Wall

The posterior abdominal wall is shorter than the anterior and lateral walls of the abdomen and is more fixed. It consists of the lumbar vertebrae and the pelvic bones. It also includes three muscles, the quadratus lumborum, the psoas major, and the iliacus. The latter two muscles flex the thigh (see Chapter 20 for more information). The quadratus lumborum and psoas major merge with the posterior margin of the diaphragm.

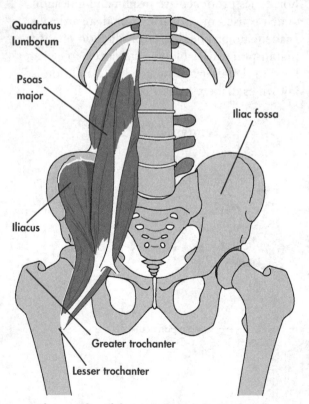

Quadratus
lumborum

Psoas
major

Iliac fossa

Iliacus

Greater trochanter

Lesser trochanter

The muscles of the posterior abdominal wall.

The posterior wall is lined by parietal peritoneum, beneath which are the abdominal aorta, the inferior vena cava, lumbar nerves, the sympathetic trunk, lymph nodes, and lymphatic vessels, along with several organs, including the kidneys, ureters, and adrenal glands.

Kidneys and Ureters: Urine Production and Transport

The kidneys are bean-shaped organs located behind the peritoneum along the upper abdominal wall. The superior pole of the left kidney lies next to the eleventh rib, while the right kidney lies next to the twelfth rib. The suprarenal or adrenal glands rest on top of each superior pole (see the next section for more on these).

The kidneys produce urine, which helps regulate the body's fluid levels and rids the blood of waste products from metabolism. They also produce hormones that keep bones strong and stimulate red blood cell production.

The hilum of each kidney contains the renal arteries and veins. From each hilus emerge the ureters that transport urine to the bladder located in the pelvis. They cross the pelvic brim in their course to the bladder. Their entrance into the bladder is discussed in Chapter 19.

Adrenal Glands: Jolting You into Action

The adrenal glands, also known as suprarenal glands, are endocrine glands that rest on top of the superior poles of each kidney. They have a different embryonic origin than the kidney, and therefore, are not part of the kidney itself. A small amount of connective tissue separates the two organs.

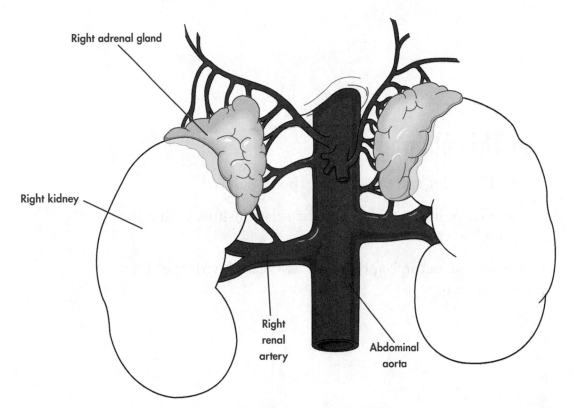

Right adrenal gland

Right kidney

Right renal artery

Abdominal aorta

The kidneys and adrenal glands.

The adrenal glands have a rich blood supply. They consist of two parts with diverse functions and embryonic origins. The outer region is called the *adrenal cortex*. It is essential for life. Its cells principally produce corticosteroid hormones, such as cortisol and aldosterone, which regulate protein and carbohydrate metabolism and sodium balance, respectively, in the body. The inner region of the adrenal glands is the *medulla*. This region is not essential for life. It consists of cells that are developmentally similar to sympathetic neurons. They produce epinephrine and norepinephrine, which amplify the effect the sympathetic nervous system elicits when it is activated in times of "fight-or-flight" stress—like when a person attempts to write a book about anatomy!

The Least You Need to Know

♦ The anterior (or front) abdominal wall is divided into quadrants and contains four muscles: external abdominal oblique, internal abdominal oblique, transversus abdominis, and rectus abdominis.

♦ The gastrointestinal (GI) tract is inside the abdominopelvic cavity and includes the esophagus, stomach, small intestine, and large intestine.

♦ The accessory organs of digestion, including the liver, gallbladder, and pancreas, are also inside the abdominopelvic cavity.

♦ The organs along the posterior (or back) abdominal wall include the kidneys and ureters, as well as the adrenal glands.

In This Chapter

◆ The inlet and outlet of the bony pelvis

◆ The pelvic organs and their relationships within the pelvic cavity

◆ The structures and organs associated with the perineum

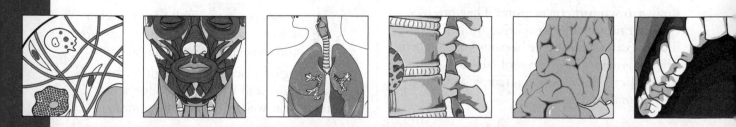

19

The Pelvis and Perineum: Stuck in the Middle of You

Elvis would have been nothing without his swiveling pelvis. Actually, neither would any of us. The pelvis serves two main purposes: to support and protect the pelvic organs, and to transfer the weight of the body to the lower limbs.

The bony pelvis is made up of the two hip bones and the sacrum. The pelvic cavity is the lower continuation of the abdominopelvic cavity and ends at the pelvic diaphragm. It contains the urinary bladder and rectum, and the respective internal reproductive organs for the two sexes. Beneath the pelvic cavity is the perineum. The perineum is the area of the body below the pelvic diaphragm and between the thighs. In this region are the external genitalia, the urethra, and the anal canal.

The pelvic region of the body is quite "busy." Structures in this area make connections with several nearby regions, such as the abdominal cavity, the gluteal region (the buttocks), the perineum, and the lower limb. It's enough to make anyone's head swivel! However, in this chapter, I will emphasize important landmarks and relationships of structures to add to the framework laid down in earlier chapters.

A Bony Bowl: The Pelvis

It's always good to start with a discussion of the bony anatomy because it represents the "walls" on which muscles, ligaments, and other structures are attached or related. This is especially true for the pelvic region.

The pelvis, meaning "basin," is formed by the right and left hip bones, which join the sacrum at the sacroiliac joints. The hip bones join one another in front at the pubic symphysis, which is a fibrocartilaginous joint. The hip bones form from three separate bones: the *ilium*, the *ischium*, and the *pubis*. These bones fuse at the *acetabulum*, the "cup" that holds the head of the femur at the hip joint. These bones are discussed further in Chapter 20.

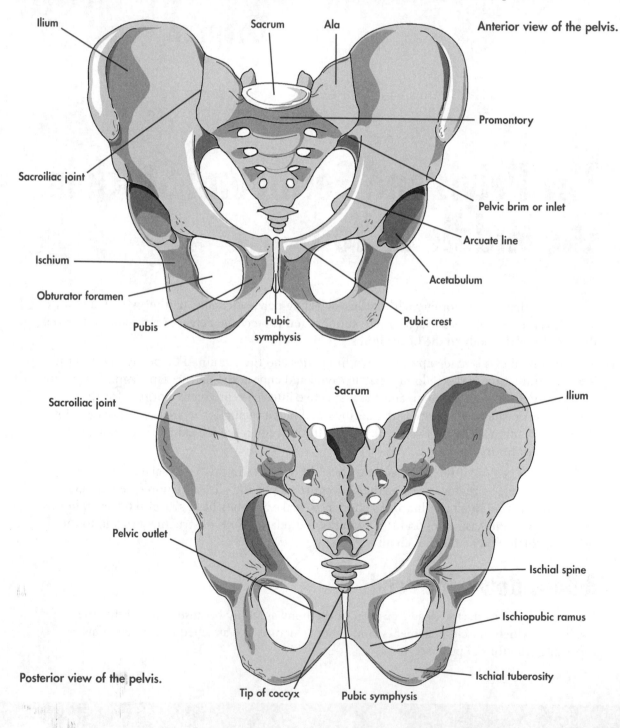

Anterior view of the pelvis.

Ilium

Sacrum

Ala

Promontory

Sacroiliac joint

Pelvic brim or inlet

Arcuate line

Ischium

Acetabulum

Obturator foramen

Pubis

Pubic symphysis

Pubic crest

Sacroiliac joint

Sacrum

Ilium

Pelvic outlet

Ischial spine

Ischiopubic ramus

Posterior view of the pelvis.

Ischial tuberosity

Tip of coccyx

Pubic symphysis

The Pelvic Openings

A pair of openings leads into and out of the pelvis:

◆ **The pelvic inlet.**   Also known as the pelvic brim, this is the upper opening. It follows a circular course involving portions of different bones. Beginning in front, a line can be drawn from the *pubic symphysis*, to the *pubic crest*, the *arcuate line* of the ilium, the *ala* and *promontory* of the sacrum, and back to the front again. The widened portion of the pelvis above the pelvic brim is called the *false pelvis*. It is the lowest part of the abdominal cavity. The narrow space below the pelvic brim is the *true pelvis*.

> **Body Language** _____
>
> The widened portion of the pelvis above the pelvic brim is called the **false pelvis**. The **true pelvis** is the area between the pelvic inlet and the pelvic outlet.

◆ **The pelvic outlet.**   A diamond-shaped area formed by bone and ligaments. In front is the pubic symphysis, on the sides are the *ischial tuberosities*, and at the back is the tip of the coccyx. Connecting the ischial tuberosities and the coccyx are thick *sacrotuberous ligaments*.

The head of an infant about to be born must be able to fit through both pelvic openings. For this reason, careful prenatal measurements of the female pelvis are made to determine whether the baby's head is too big to fit through these openings.

You Floor Me: The Muscular Pelvic Diaphragm

The pelvic organs rest on a funnel-shaped muscular floor called the *pelvic diaphragm*. This thin diaphragm consists of skeletal muscles that attach to the inner walls of the pelvic cavity from the pubic symphysis to the coccyx. The muscles that make up the pelvic diaphragm are …

◆ **The levator ani.**   This muscle group consists of the puborectalis, pubococcygeus, and iliococcygeus muscles.

◆ **The coccygeus.**   They are located behind the levator ani. They attach from the ischial spines to the sides of the lower sacrum and coccyx.

There is a gap (the *urogenital hiatus* or opening) between the right and left levator ani through which pass the urethra, vagina, and anal canal in the female, and the urethra and anal canal in the male. In reality, the adjacent muscular edges of the levator ani attach to the outer walls of these midline organs and seal this space completely. Collectively, these muscles support the pelvic organs.

Lined Up in the Midline: The Pelvic Organs

The pelvic organs are located in the midline of the pelvic cavity. In both males and females, the urinary bladder is located at the front of the cavity, behind the pubic bone. The rectum is located toward the back, in front of the sacrum. Between these, in the male, are the ductus deferens, seminal vesicles, and the prostate gland. In the female are the uterus, vagina, uterine tubes, and ovaries.

Ureters and Urinary Bladder

The ureters pass across the common iliac arteries just as they divide into the internal and external iliac arteries at the pelvic brim and take a course along the lateral pelvic wall before making their final approach to the lower back wall of the bladder.

◆ **In the female.** The ureter is crossed over near the level of the cervix by the uterine artery. This so-called "water (ureter) under the bridge (artery)" relationship is important. During surgical removal of the uterus (a hysterectomy), the ureter cannot be clamped or tied off with the artery; it would be injured.

◆ **In the male.** The ductus deferens crosses over the ureter near its entrance into the bladder. The ureters pass through the wall of the bladder diagonally and have slit-like openings on its internal surface.

Bodily Malfunctions

Occasionally, a ureter may become blocked, for example, by a kidney stone. This may occur anywhere along its course. One frequent site is where the ureter enters the bladder wall. Blockage causes the urine to "dam up" above the site of blockage, causing stretching and enlargement of the ureter or calyces within the kidney. This condition is known as hydronephrosis. If left untreated, it may lead to loss of kidney function.

The urinary bladder is a hollow, muscular sac-like organ located beneath the peritoneum behind the pubic bones. It is a passive recipient of the urine continuously arriving from the ureters. Once the bladder fills to a certain size, nerves within its walls inform the central nervous system that it is getting full and a person feels the urge to urinate. The smooth muscle wall of the bladder is the *detrusor muscle*. The internal mucosal lining over most of the bladder is folded and smoothes out as the bladder fills.

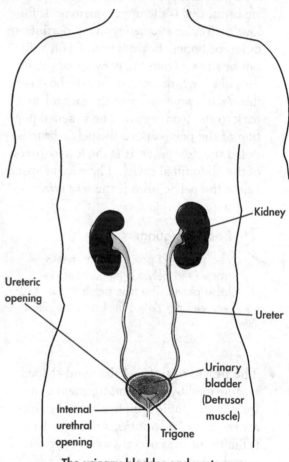

The urinary bladder and ureters.

At the back wall of the bladder is a small triangular area called the *trigone*. Here, the mucous membrane is smooth and tightly attached to the muscle. At the upper outer "corners" of this triangle are the openings of the ureters. The bottom corner is the opening to the urethra. Smooth muscle around this internal urethral opening forms a physiologic sphincter, called the *internal urethral sphincter*. It keeps urine from entering the urethra. In males, it keeps semen from entering the bladder during ejaculation.

Foot Notes

The diagonal course of the ureters through the bladder wall and their narrow openings inside the bladder serve a valve-like mechanism to keep urine from going back up the ureters when the bladder is full and during urination when the bladder contracts. Failure of this mechanism can also lead to hydronephrosis.

The empty bladder is a pelvic organ. When it fills, it may rise above the pubic bones into the lower anterior abdominal wall. Its location behind the pubic bones means that it is vulnerable to injury following pelvic fractures, for example, in a car accident.

Rectum

The sigmoid colon crosses the pelvic brim and enters the pelvic cavity. It ends in front of the sacrum as the rectum. The rectum continues through the pelvis and ends at the anal canal, where it passes through the pelvic diaphragm to enter the perineum. The rectum, like the bladder, is a storage organ. It accumulates and holds the fecal mass until it passes out the anal canal and anus during defecation.

Male Pelvic Organs

The male pelvic organs include the prostate gland, the ductus deferens, and the seminal vesicles.

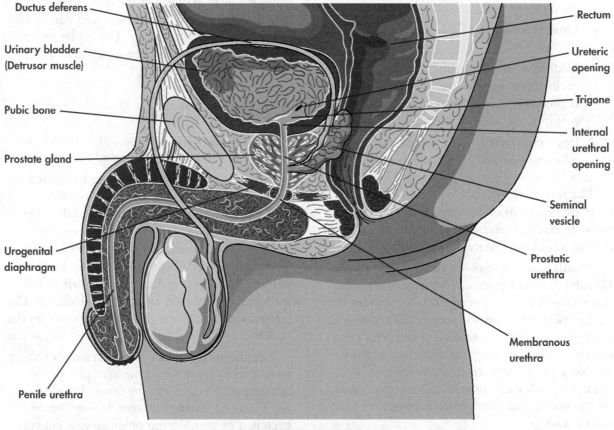

Ductus deferens

Urinary bladder
(Detrusor muscle)

Pubic bone

Prostate gland

Urogenital
diaphragm

Penile urethra

Rectum

Ureteric
opening

Trigone

Internal
urethral
opening

Seminal
vesicle

Prostatic
urethra

Membranous
urethra

A sagittal view of the male pelvic cavity.

The prostate gland is a walnut-sized fibro-muscular organ that rests on the pelvic diaphragm and urogenital hiatus. Its function was discussed in Chapter 12. The urethra passes through the center of the prostate after leaving the bladder. This portion of the urethra is the *prostatic urethra*. Projecting into the posterior surface of the prostatic urethra is a midline ridge, called the *urethral crest*. This is formed by one of the lobes of the prostate, the median lobe. Into the urethral crest open the ejaculatory ducts. On either side of this crest are several tiny openings for prostatic ducts. Prostatic secretions are added to the semen through these ducts during ejaculation.

The prostate gland has five lobes:

♦ **Anterior lobe.** Lies in front of the urethra. It contains few glands and is not a common site of disease.

♦ **Lateral lobes (2).** Lie along the sides of the urethra.

♦ **Middle or median lobe.** Located in the midline just behind the prostatic urethra. It produces the urethral crest. It also projects into the neck of the bladder.

♦ **Posterior lobe.** Lies behind the median lobe and contains numerous glands.

The posterior lobe is a common site of prostate cancer in older men. This part of the prostate can be readily felt through the rectum during a routine physical examination.

The ductus deferens is a thick-walled muscular tube that transports sperm from the epididymis to the prostatic urethra. It is a component of the spermatic cord and enters the abdominal cavity via the inguinal canal and the deep inguinal ring. After entering the abdomen, it passes medially and downward to enter the pelvis. Near the back wall of the bladder, it is crossed by the ureter and then passes between the seminal vesicles.

Bodily Malfunctions

Noncancerous enlargement of the median lobe, called benign prostatic hypertrophy (BPH), also commonly occurs in men over 50. In this latter condition, enlargement of the median lobe pushes forward into the urethra or neck of the bladder and may cause partial or complete obstruction. This interferes with urine flow, causes an increased urgency and frequency of urination (the "growing/going problem" they talk about in Flomax ads), and causes retention of urine in the bladder.

The *seminal vesicles* are paired, coiled tubes located on the backside of the bladder. They produce an "energy cocktail" containing fructose for sperm arriving in the ductus deferens. The ducts of the seminal vesicle and ductus join to form the *ejaculatory ducts*. These pass through the prostate gland and open at the urethral crest of the prostatic urethra.

Female Pelvic Organs

The female pelvic organs consist of the uterus, the vagina, the uterine tubes, and the ovaries.

The uterus is a pear-shaped organ located in the midline of the pelvic cavity. It consists of a *fundus*, located above the entrance of the uterine tubes; the *body* or main part of the organ; and the *cervix*, which projects into the vagina. The circular space between the cervix and the walls of the vagina is divided into parts called the fornices: anterior, lateral, and posterior. The uterus is normally bent forward and rests on the bladder.

The uterus is held in place and supported by several ligaments and the muscular pelvic diaphragm. These form the floor of the pelvis. These ligaments and muscle may become stretched or torn during pregnancy or delivery. With advancing age and menopause, they lose

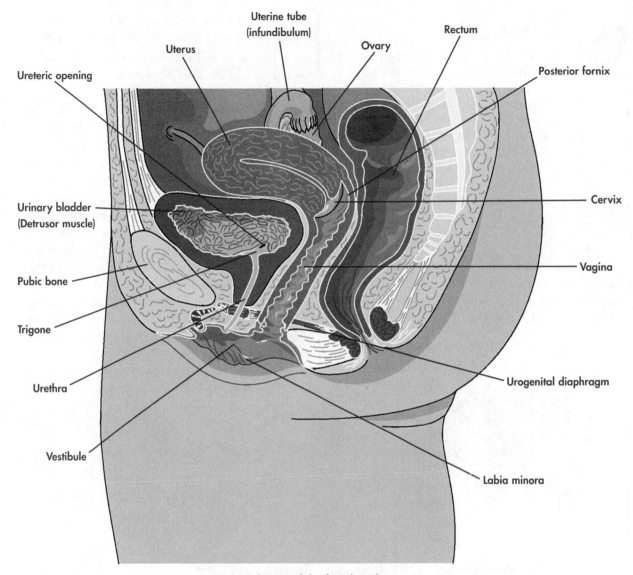

A sagittal view of the female pelvic cavity.

their ability to support the uterus. When this occurs, the uterus may sag into the vagina, a condition known as uterine prolapse.

The vagina is a stretchy fibromuscular tube that attaches to the uterus at the cervix. It opens into the vestibule, between the labia minora. Its front wall is shorter than its back wall. The urethra is located against the anterior wall of the vagina in its course from the bladder to the vestibule.

The uterine tubes extend laterally from the uterus to the ovaries. Near the ovary, they broaden to a funnel-shaped end, called the *infundibulum*, which is surrounded by the *fimbriae*. The uterine tube is hollow and connects the uterine cavity (and vagina) with the peritoneal cavity. Eggs released monthly by the ovaries enter the uterine tubes and are transported to the uterus.

Foot Notes

The fact that the uterine and peritoneal cavities are continuous with one
another is used clinically to determine
whether blockage exists along the course
of one or both uterine tubes in women who
are having difficulties becoming pregnant. A
doctor injects dye into the uterus through the
vagina. It flows up through the uterine tubes
and into the peritoneal cavity. If there is a
blockage in the tubes, it shows up on an x-
ray where the dye stopped.

The ovaries are almond-shaped structures
located near the side walls of the pelvis. They
are held in place by several structures. These
include their blood supply, the ovarian ligament

(which attaches it to the side of the uterus), and
the mesovarium, a part of the broad ligament.

The uterus and uterine tubes are enclosed
within a double layer of peritoneum that
extends sideways to the lateral walls of the pelvis. This "broad" ligament is, in fact, called the
broad ligament. The front part of this ligament
continues onto the upper surface of the bladder and the anterior abdominal wall. The back
part of this ligament is lifted onto the rectum
and continues upward on the posterior wall of
the pelvis. The part of the broad ligament that
encloses the uterine tubes is called the *mesosalpinx* (-salpinx refers to "tube"). The largest
part that extends laterally from the sides of the
uterus is the *mesometrium.* The ovaries attach
to the back surface of the broad ligament by a
shelf-like fold called the *mesovarium.* The broad
ligament provides additional structural support
for the uterus.

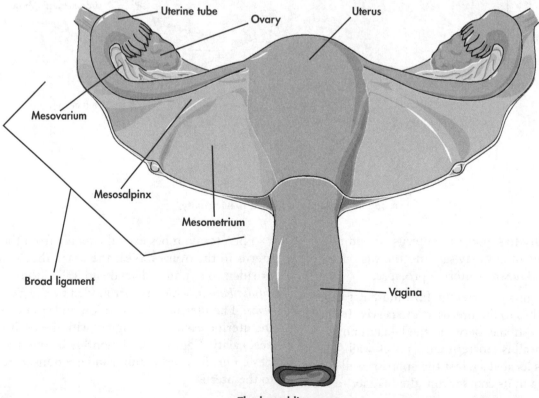

The broad ligament.

A Diamond in the Rough: The Perineum

Below the pelvic diaphragm is the perineum. The perineum is a diamond-shaped area between the thighs that contains the external genitalia and the anal canal and anus. Its boundaries are the pubic symphysis in front, the ischial tuberosities of the ischium bones laterally, and the tip of the coccyx in back.

What do you get when you divide a diamond-shaped figure in half? Two triangles, right? And so, indeed, the perineum is described as having two different triangles that have different structures associated with them. A line connecting the two ischial tuberosities creates the urogenital triangle (in front) and the anal triangle (in back).

The Urogenital (UG) Triangle

The urogenital triangle and region contains the external genitalia and the urethra. The boundaries of this triangle are the ischiopubic rami (fused portions of the pubis and ischium bones) and the line between the ischial tuberosities. This triangular space is filled by the urogenital diaphragm, which is a thin skeletal muscle sheet enclosed by connective tissue.

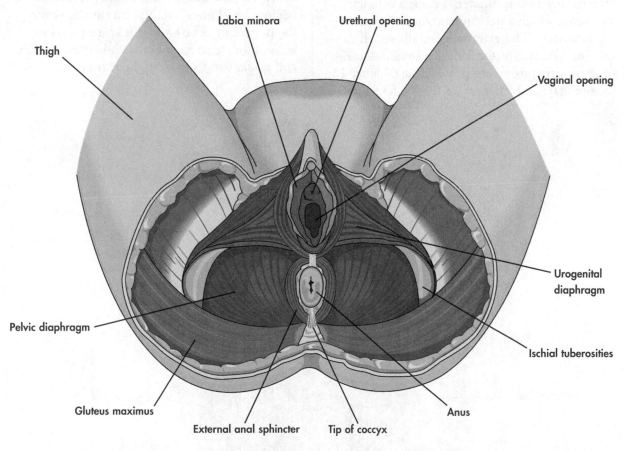

Thigh

Labia minora

Urethral opening

Vaginal opening

Urogenital diaphragm

Pelvic diaphragm

Ischial tuberosities

Gluteus maximus

External anal sphincter

Tip of coccyx

Anus

The female perineum.

The UG diaphragm is located just beneath the urogenital opening of the pelvic diaphragm. After passing through this opening, the urethra and vagina in the female pass through the UG diaphragm to enter the vestibule. In the male, only the urethra passes through the urogenital opening and the UG diaphragm. The portion of the urethra that passes through the UG diaphragm is the membranous urethra. This is the shortest portion of the urethra. In the male, it connects the prostatic urethra with the penile urethra, which opens externally at the glans penis.

Muscle within the UG diaphragm of both sexes surrounds the urethra and forms the external urethral sphincter. This is a voluntary sphincter, which a person has control over during urination. The external genitalia attach to the undersurface of the UG diaphragm. Details of these structures were provided in Chapter 12 for the male and Chapter 13 for the female.

The Anal Triangle

The anal triangle and region contains the anal canal. Considerable amounts of fat surround it. This triangle is bordered in front by the common line connecting the ischial tuberosities, along the sides by the sacrotuberous ligament and lower edge of the gluteus maximus (a muscle of the buttocks), and behind by the tip of the coccyx. The coccyx is the apex of the anal triangle. The "roof" of this wedge-shaped space is the undersurface of the pelvic diaphragm. The rectum passes through the urogenital hiatus. Here, its back wall is looped and bent forward by the puborectalis muscles to form the anorectal junction. This is where the anal canal begins. The "floor" of this space is the skin of the perineum. The fat in this space yields to accommodate passage of the child during birth and passage of feces from the rectum.

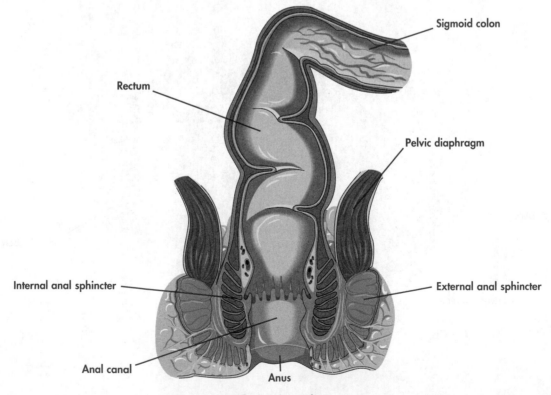

Sigmoid colon

Rectum

Pelvic diaphragm

Internal anal sphincter

External anal sphincter

Anal canal

Anus

The anal canal.

The anal canal is about an inch and a half long and ends at the anus. It is controlled by two sets of sphincters. The internal anal sphincter is a thickening of circular smooth muscle in the wall of the canal. It is under involuntary control by the autonomic nervous system. The external anal sphincter, on the other hand, is composed of skeletal muscle and is under a person's voluntary control by somatic nerves. Injury to these muscles or to their nerve supply can cause fecal incontinence.

Bodily Malfunctions

Occasionally, the veins around the anal canal become enlarged and inflamed. This common condition is known as hemorrhoids. Hemorrhoids may result from constipation and straining to defecate. They are also common among pregnant women. In both cases, increased pressure on the veins in this area interferes with their venous return, causing them to enlarge.

The Least You Need to Know

◆ The bony pelvis protects the pelvic organs and transfers the weight of the body to the lower limbs.

◆ The floor of the abdominopelvic cavity is the pelvic diaphragm, consisting of the levator ani and coccygeus muscles. It provides support for the pelvic organs.

◆ Below the pelvic diaphragm is the perineum. The perineum is a diamond-shaped area between the upper thighs and contains the external genitalia and the anal canal.

In This Chapter

- ◆ The regional parts and bones of the lower limb
- ◆ The muscle compartments in the lower limb and their principal actions
- ◆ The basic nerve and blood supply to the lower limb

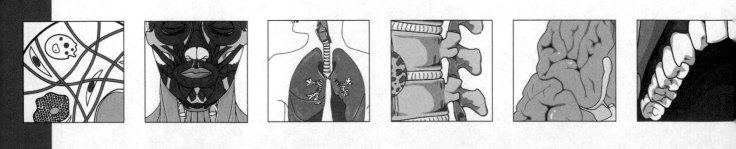

Chapter 20

The Lower Limb: From the Waist Down

The late comedian Rodney Dangerfield's most famous line, "I don't get no respect," could apply to the lower limb. There's really nothing fancy about it. It consists of the hip, thigh, leg, and foot (on both sides). It supports the weight of the body and serves as transportation. We use our lower limbs to get us from point A to point B. On the other hand, though, for professional soccer players and NFL place kickers, they serve as "pay checks," which simply means that the average person doesn't develop their lower limbs to their full potential.

Learning about the lower limb is easier because you will see many similarities with the organization of the upper limb, described in Chapter 16. In this chapter, I attempt to minimize the detailed anatomy of this region by focusing on the basic features of bones, muscle compartments, and the nerve and blood supply of the lower limb. In the end, you will gain an understanding of how this marvelous appendage works and some common problems it encounters during a person's life.

Dem Bones, Dem Bones: Regions and Bones of the Lower Limb

The lower limb consists of the hip, thigh, leg, and foot. The hip consists of the hip bone, or *os coxae*, which attaches to the femur. Here, the weight of the body's trunk is shifted onto the lower limb. The thigh is the region of the lower limb between the hip and the knee. It is analogous to the arm in the upper limb. It contains a single bone, the *femur*. The leg is that region of the lower limb between the knee and the ankle. Like the forearm, it contains two bones, the tibia and the fibula. The foot is the terminal end of the lower limb. Here, the weight of the body is transmitted to the ground. There are 26 bones of the foot, including seven tarsal bones, five metatarsal bones, and 14 phalanges. I give details of each region in the next sections.

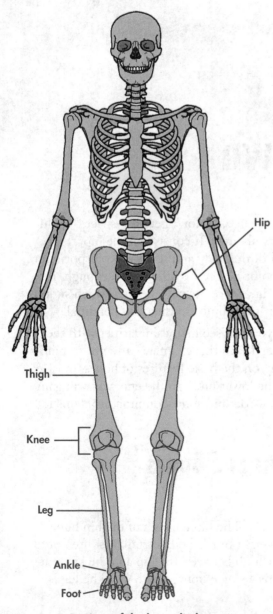

Regions of the lower limb.

Butt Out!: The Hip and Gluteal Region

The hip bone is a single bone in an adult, formed by the union of three bones: the *ilium*, the *ischium*, and the *pubis*. Portions of these bones fuse at puberty at the *acetabulum*, the socket for the head of the femur. The pelvis is the bony structure formed by union of the hip bones with the sacrum and the pubic bones with the pubic symphysis (see Chapter 19 for more about the pelvis). Two kinds of joints are formed at these sites of union. The *sacroiliac* joints are synovial joints between the sacrum and the ilium portion of the hip bone. These joints are strengthened by strong ligaments. These joints don't move much, though. The pubic symphysis is a fibrocartilaginous joint that is also reinforced by ligaments. Little movement is permitted at this joint as well. These joints strongly connect the pelvis to the axial skeleton and provide a stable frame to which the lower limb attaches and moves. Details of specific parts of these bones are in the accompanying figures.

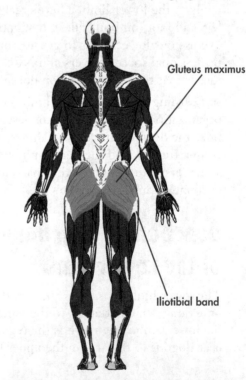

Muscles of the gluteal region, superficial layer.

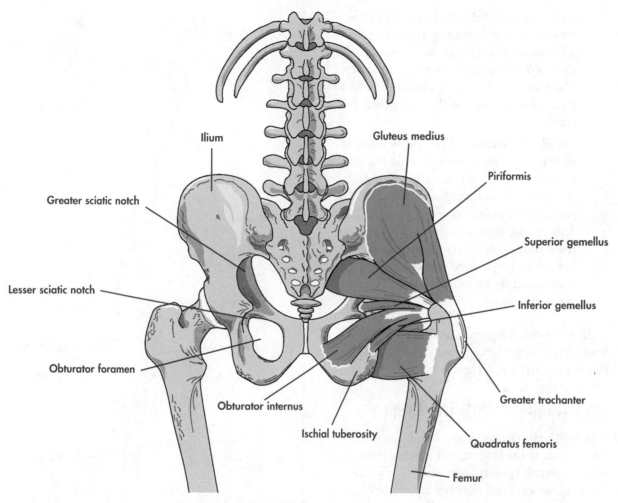

Muscles of the gluteal region, deep layer.

The gluteal region (also known as the buttocks) is located behind the hip bone. It contains the gluteal muscles and several smaller muscles. These muscles are the following:

♦ **Gluteus maximus.** As its name suggests, it is the largest of the gluteal muscles. It covers most of the other muscles of this region. The gluteus maximus is the most powerful extensor of the hip. But it is used only when maximal exertion is required, for example, when we get up out of a chair or when we walk up a steep flight of steps. The inferior gluteal nerve is the motor nerve supply to the gluteus maximus.

♦ **Tensor fasciae latae.** The other superficial muscle of the gluteal region. It is located on the outside surface of the ilium. It is enclosed within a thickening of the fascia lata of the thigh, called the *iliotibial band*. This muscle's principal action is flexion of the hip.

♦ **Gluteus medius and gluteus minimus.** These muscles lie beneath the maximus and are the principal abductors and medial rotators of the thigh, movements that carry the limb away from the side of the body or cause the big toe to be pointed inward, respectively. They attach from the outer surface of the ilium to the femur. However, their main function is to keep the pelvis stable and level when weight is taken off the limb, as during walking or standing on one leg. The gluteus medius and minimus and the tensor fasciae latae are innervated by the superior gluteal nerve.

◆ **Smaller muscles.** Located beneath the gluteus maximus, these include the piriformis, obturator internus, superior and inferior gemelli, and quadratus femoris. They connect the pelvis to an area on the femur. As a group, these muscles laterally rotate the femur, a movement that causes the big toe to be pointed outward. They are innervated by small nerve branches from the sacral plexus (S1–S3).

In addition, the principal nerve supply of the lower limb, the sciatic nerve, passes through this region from its origin within the pelvis.

Let's Thigh One On!: The Thigh

The thigh contains a single bone, the femur. The femur is the largest and longest bone of the body. Its principal features include a large, spherical head that fits snugly into the acetabulum (hip socket), a cylindrical neck that angles downward to attach to the shaft, and a long shaft that enlarges at the knee and ends at two round prominences, the *lateral and medial condyles*. Where the neck unites with the shaft are two large bony projections for attachment of muscles. They are the *greater and lesser trochanters*. Both the femoral head and the condyles are covered with articular cartilage. Between the femoral condyles is a depression, the *intercondylar fossa*, for attachment of the cruciate ligaments of the knee.

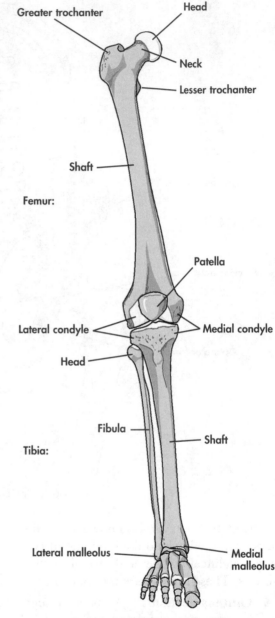

Bones of the lower limb.

Bodily Malfunctions

The neck of the femur is a common site of fracture, particularly in elderly postmenopausal women whose bones may have become weakened by osteoporosis. Falls are the common cause of these fractures. These fractures usually disrupt the blood supply to the femoral head and it dies. A hip replacement operation is usually performed to treat this condition.

The muscles of the thigh are some of the largest muscles in the body. These are contained beneath the skin and superficial fascia in a thick, fibrous connective tissue wrapping of deep fascia, called the *fascia lata* (but don't try to order it at the coffee shop). The lateral portion of this fascia is thickened as the *iliotibial band*, which attaches to the upper outside surface of the tibia.

Membrane sheets from the fascia lata penetrate into the thigh and attach to the femur, separating the muscles of the thigh into compartments, similar to the organizational plan in the upper limb (see Chapter 16). The thigh has three muscle compartments that formed during embryologic development: anterior compartment, medial compartment, and posterior compartment. Muscles within each of these compartments generally have a common function and nerve supply, as follows:

◆ **Anterior compartment.** Muscles in this compartment include the sartorius, the quadriceps femoris (which consists of the vastus lateralis, the vastus intermedius, the vastus medialis, and the rectus femoris), and the iliopsoas. The iliopsoas, sartorius, and rectus femoris flex the thigh, whereas

the remaining muscles of the quadriceps femoris extend the leg. They all get their nerve supply from the femoral nerve (L2, 3, 4).

◆ **Medial compartment.** Muscles in this compartment are the adductor longus, adductor brevis, adductor magnus, gracilis, and pectineus. Their job, as you might guess, is to adduct and flex the thigh. They get their nerve supply from the obturator nerve (L2, 3, 4).

◆ **Posterior compartment.** Muscles in this compartment are the biceps femoris, semitendinosus, and semimembranosus. They extend the thigh and flex the leg. They get their nerve supply from the sciatic nerve (L4–S3).

The principal muscles of the anterior compartment include the quadriceps femoris and the iliopsoas muscles. The quadriceps femoris is a large muscle consisting of four parts. The vastus muscles have an extensive attachment to the shaft of the femur. Tendons from these muscles combine with the tendon of the rectus femoris to form the quadriceps tendon that attaches to and contains the patella (kneecap bone). The patella attaches to the tibial tuberosity via the patellar ligament. The quadriceps muscle is the principal extensor of the leg.

Foot Notes

The patella is embedded within the quadriceps tendon. It is an example of a sesamoid bone, a bone contained within a tendon that crosses a joint. The patella is the largest sesamoid bone in the body. Such bones help increase the strength of muscle pull across a joint.

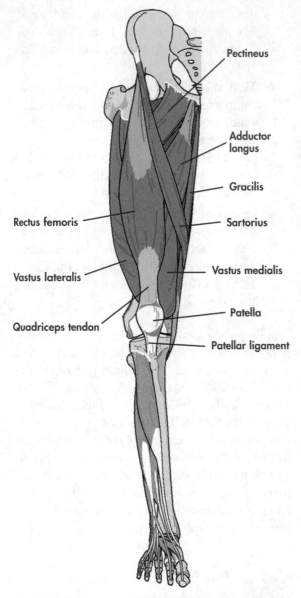

Pectineus

Adductor longus

Gracilis

Rectus femoris

Sartorius

Vastus lateralis

Vastus medialis

Quadriceps tendon

Patella

Patellar ligament

Muscles of the anterior and medial compartments of the thigh.

The principal flexor of the thigh is the ilio-psoas muscle, which is formed by the union of the psoas major and iliacus muscles that originate in the abdomen. They combine and pass into the thigh beneath the inguinal ligament and attach to the lesser trochanter of the femur.

The muscles of the medial compartment are the adductors of the thigh. Their action is to bring the limb closer to the body. For example, they prevent the limb from looping too far out during walking as the unweighted limb is swung forward.

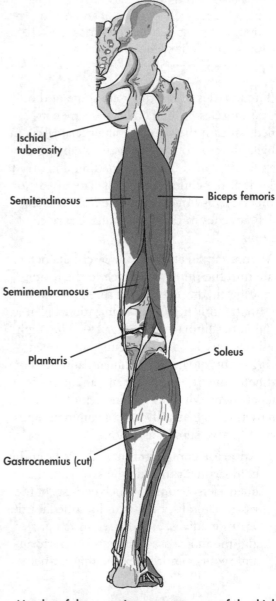

Ischial tuberosity

Semitendinosus

Biceps femoris

Semimembranosus

Plantaris

Soleus

Gastrocnemius (cut)

Muscles of the posterior compartments of the thigh and leg.

The posterior compartment muscles, biceps femoris, semitendinosus, and semimembranosus, are referred to as the hamstrings. These are powerful muscles that extend the thigh and flex the leg. They are well-developed in track and field athletes and soccer players, who have to accelerate with bursts of speed.

Bodily Malfunctions

A pulled hamstring is a sudden, painful injury that involves stretching or tearing a hamstring muscle where it attaches to the ischial tuberosity. Depending on the severity of the injury, surgery may be needed to reattach the muscle to the bone.

Getting a Leg Up on the Competition: The Leg

The leg contains two bones, the tibia and the fibula. The tibia, also known as the shin bone, is the larger of the two. It bears the weight of the leg and is located on the inside (medial side). Its proximal portion is broad where it articulates with the femur and bears the two flat, smooth medial and lateral condyles. Between the condyles is a bony ridge called the *intercondylar eminence*. This is where the cruciate ligaments of the knee attach. In addition, just below the condyles at the front of the tibia is a large roughened bony prominence called the *tibial tuberosity*. This is easily felt and seen through the skin. The strong, thick patellar ligament attaches here.

Bodily Malfunctions

One of the most common "diseases" of young athletes going through the growth spurt is Osgood-Schlatter's disease. This is not really a disease, but an overuse injury that involves the tibial tuberosity where the patellar ligament attaches. Stress on this actively growing tuberosity by the quadriceps muscles causes mild to severe pain during running or jumping. The pain usually goes away once the bone stops growing.

Continuing downward from the tibial tuberosity is the sharp anterior border of the bone, which is covered only by skin and also readily felt. This area of the bone is the "shin," which hurts when it is hit. The tibia ends at the ankle, where it articulates with the talus bone at the ankle or *talocrural joint* ("crura" means leg). A bony process called the *medial malleolus* projects downward on the inside of the tibia. That is the prominent "bump" that you can see through the skin and feel on the inside of the ankle.

The fibula is the long, slender, nonweight-bearing bone on the outer side of the leg. It has a small, round head that attaches to the upper outer surface of the tibia. Its lower end enlarges and also participates in the ankle joint. It is the prominent bump on the outside of the ankle. The shafts of the fibula and the tibia are further connected to each other by an interosseous membrane, which muscles attach to.

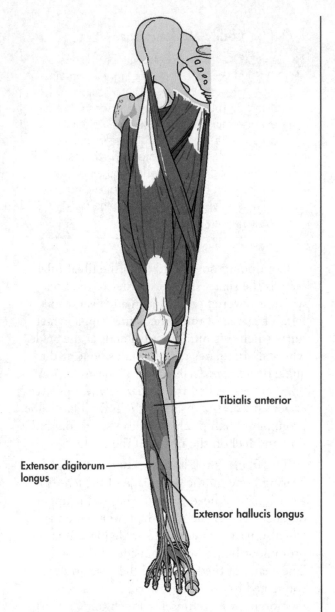

Muscles of the anterior compartment of the leg.

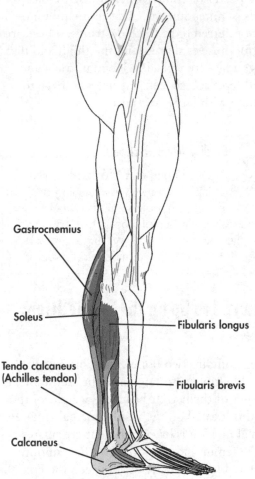

Muscles of the lateral and posterior compartments of the leg.

The fascia lata of the thigh continues in the leg as the *crural fascia*. This fascia also subdivides the leg into functional compartments:

◆ **Anterior compartment.** Located on the front of the leg between the tibia and fibula. The muscles of this compartment are the tibialis anterior (dorsiflexes the ankle and inverts the foot), extensor digitorum longus (extends the lateral four toes and dorsiflexes the ankle), extensor hallucis longus (extends the big toe and dorsiflexes the ankle), and fibularis tertius (dorsiflexes the ankle and everts the foot).

◆ **Lateral compartment.**   Confined to the area on the outside (lateral) surface of the fibula. Muscles are the fibularis longus and fibularis brevis. They evert the foot and plantar flex the ankle.

◆ **Posterior compartment.**   Located on the back of the leg between the tibia and fibula. This compartment contains two muscle groups. The superficial group (the "calf" muscles) consists of the gastrocnemius, soleus, and plantaris. They plantar flex the foot. The deep group consists of the popliteus (unlocks the knee), flexor digitorum longus (flexes the lateral four toes), flexor hallucis longus (flexes the big toe), and tibialis posterior (plantar flexes and inverts the foot).

The Ankle and Foot: Carry That Weight

When a person stands, the foot makes a right angle to the leg and transmits the weight of the tibia and body to the ground. The bones of the foot include the following:

◆ **Seven tarsal bones.**   Include the talus, calcaneus (the heel bone), navicular, medial cuneiform, intermediate cuneiform, lateral cuneiform, and cuboid.

◆ **Five metatarsal bones.**   Connect the toes to the rest of the foot.

◆ **Fourteen phalanges.**   The big toe has two phalanges, whereas the other toes have three.

The talocrural (ankle) joint is a hinge-type synovial joint between the distal ends of the tibia and fibula and the talus bones. Dorsiflexion and plantar flexion occur at this joint. The joint capsule is thin and loose. It is reinforced on its medial and lateral sides by collateral ligaments. The medial collateral ligaments resist eversion of the foot. The lateral collateral ligaments resist inversion of the foot. The lateral collateral ligaments are weaker than the medial collateral ligaments and are frequently stretched or torn in sprains of the ankle.

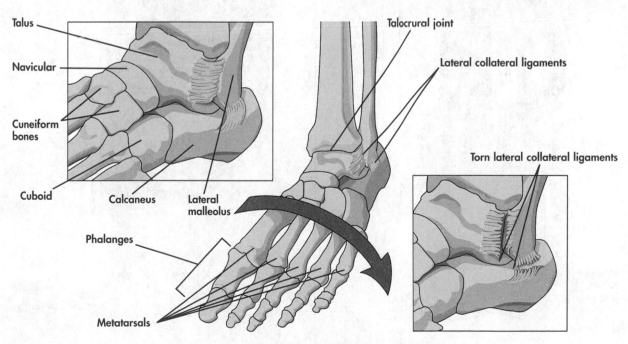

Bones of the foot and the lateral collateral ligament.

The foot bears the weight of the upright posture. It has two principal surfaces, the dorsum (top) of the foot, and the sole (bottom) of the foot. The top of the foot is equivalent to the back of the hand in that the skin is thin and loose. Through it you can see tendons coursing to the toes. The tendons on the dorsum of the foot are mainly the extensor digitorum longus (to the lateral four toes) and the extensor hallucis longus (to the big toe).

The sole or plantar surface of the foot consists of thick skin attached to an equally thick layer of superficial fascia. This contains tough fibrous connective tissue and fat. This material is bound to the plantar aponeurosis, which is the thickened central portion of deep fascia of the foot that extends from the heel to the toes. It functions as a shock-absorber to cushion the sole of the foot and support the foot arches.

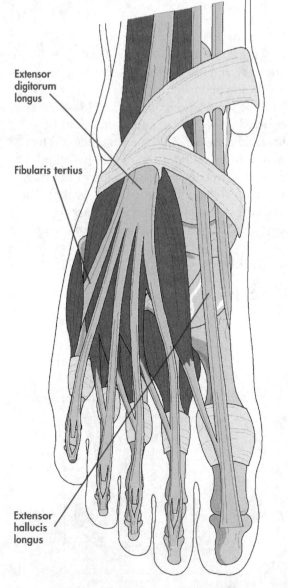

Tendons of muscles on dorsum of foot.

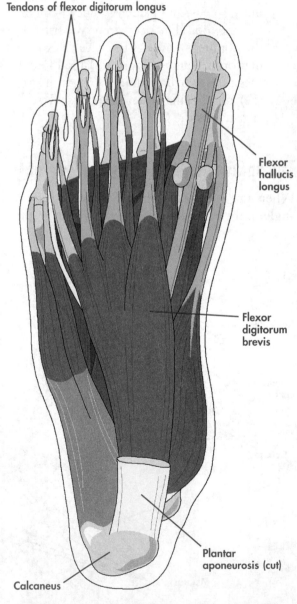

Tendons and muscles on plantar surface of the foot.

Bodily Malfunctions

The plantar aponeurosis may become stretched or torn from running or prolonged standing, resulting in inflammation and swelling. This produces tenderness and pain that extends from the heel along the inner side of the foot. This is known as plantar fasciitis.

Deep to the plantar aponeurosis are several layers of muscles and tendons of extrinsic and intrinsic muscles, similar to the organization in the hand (see Chapter 16).

I'm Hip to That: The Hip Joint

Unlike the shoulder joint, which is very mobile but less stable, the hip joint is constructed for maximal stability over mobility. The hip joint is a ball-and-socket type joint. The bony components of the joint are the acetabulum and the head of the femur. The acetabulum is a deep, cup-like depression lined by a horseshoe-shaped piece of articular cartilage. Its margin is raised by a "lip" of fibrocartilage to deepen the socket. The head of the femur fits into the acetabulum.

A strong, fibrous capsule surrounds the hip joint and attaches from the acetabulum to the neck of the femur near the trochanters. The capsule is strengthened by tough ligaments whose connective tissue fibers become tight when the limb is extended and lax when the limb is flexed. The joint is most stable when the limb is extended (when a person is standing up) and most unstable when the limb is flexed (when a person is sitting down). For this reason, it is common for the hip to be dislocated posteriorly in a car accident when the knees hit the dashboard.

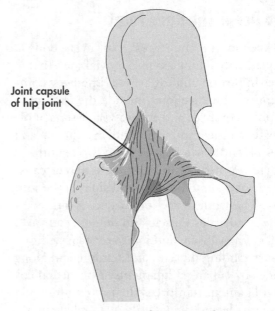

Joint capsule of hip joint

Hip joint and joint capsule.

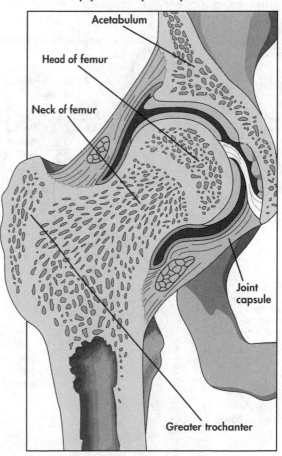

Acetabulum

Head of femur

Neck of femur

Joint capsule

Greater trochanter

Inside of hip joint.

Knee Deep: The Knee Joint

The knee joint is the largest joint in the body. It is primarily a hinge-type synovial joint that moves in flexion and extension. The bony components of the knee joint include the femoral and tibial condyles and the posterior surface of the patella (kneecap), which fits in a groove and glides up and down between the two femoral condyles. A more or less complete fibrous capsule encloses only the back and sides of the joint cavity. The front of the joint cavity is formed by the quadriceps muscle and tendon, and the patella. The capsule and joint are stabilized by the surrounding muscles and tendons and along its sides by collateral ligaments. The medial collateral ligament reinforces the joint capsule on its inner side and the lateral collateral ligament reinforces the capsule on its outer side.

Sandwiched between the femoral and tibial condyles within the joint cavity are wedge-shaped pieces of fibrocartilage called the *menisci*. These deepen or improve the fit between the condyles, which otherwise is quite shallow. Between the lateral condyles is the lateral meniscus. Between the medial condyles is the medial meniscus. The lateral meniscus is O-shaped, whereas the medial meniscus is C-shaped. In addition to deepening the contact between the condyles, the menisci also function as "shock absorbers" to reduce pounding across these bones and as "wicks" to distribute synovial fluid within the joint space. The lateral meniscus does not attach to the lateral collateral ligament. Therefore, it is more mobile within the joint cavity. The medial meniscus, on the other hand, is firmly attached to the medial collateral ligament, making it more vulnerable to injury.

Ligaments inside the joint cavity also contribute significantly to joint stability. These are the *anterior and posterior cruciate ligaments*. They are thick and strong and cross each other like an X (cruciate means "cross"). These ligaments resist excessive forward and backward movement of the femur on the tibia.

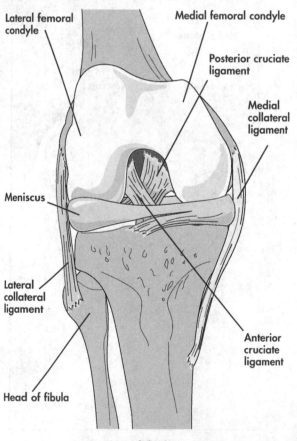

Front view of the knee joint.

Lateral femoral condyle

Medial femoral condyle

Posterior cruciate ligament

Medial collateral ligament

Meniscus

Lateral collateral ligament

Head of fibula

Anterior cruciate ligament

Bodily Malfunctions

Knee injuries are common in contact sports, such as football. A hit on the outside of the knee may cause tearing of the medial collateral ligament, the medial meniscus, and the anterior cruciate ligament. An injury to these three structures is called "the unhappy triad of the knee."

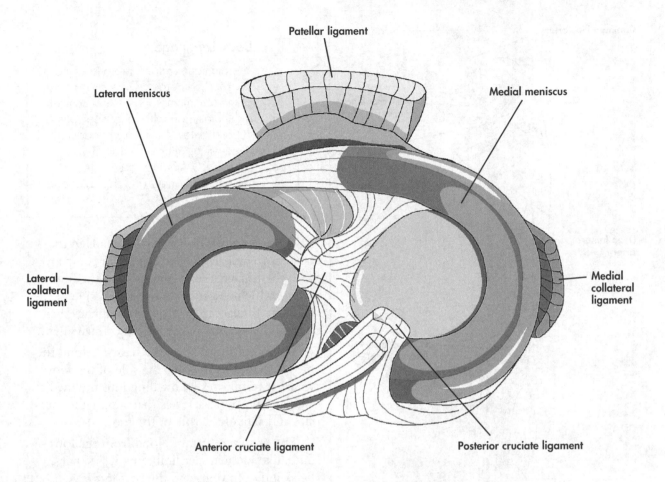

Patellar ligament

Lateral meniscus

Medial meniscus

Lateral collateral ligament

Medial collateral ligament

Anterior cruciate ligament

Posterior cruciate ligament

View of the left knee joint from above.

Blood Supply to the Lower Limb

The blood supply to the lower limb is provided by the *femoral artery*. This vessel is the continuation of the external iliac artery that originates in the abdominal cavity. It enters the thigh by passing under the inguinal ligament. Shortly after entering the thigh, it divides into the *deep femoral artery*. The deep femoral artery gives rise to the *lateral and medial circumflex femoral arteries* that provide branches that supply blood to the head and neck of the femur and overlying muscles and skin. It also gives rise to several branches that enter and supply the posterior compartment muscles of the thigh.

The femoral artery continues in a downward and medial direction in the thigh and then passes behind the knee, where its name changes to the *popliteal artery*. This artery provides several branches to the knee joint before dividing into the anterior and posterior tibial arteries. The anterior tibial artery courses over the interosseous membrane to enter and supply the anterior compartment of the leg. In the lower part of the leg, it courses in front of the ankle onto the dorsal surface of the foot and becomes the *dorsalis pedis* artery. Here, it supplies branches to the ankle and foot and a "pedal" pulse can be felt.

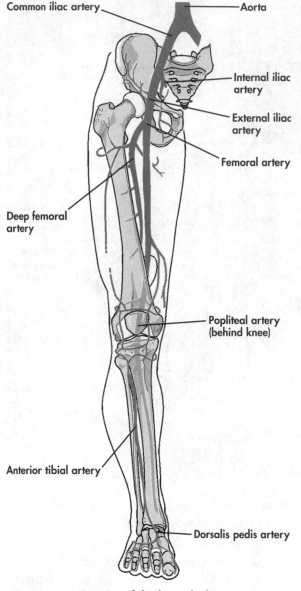

Common iliac artery

Aorta

Internal iliac artery

External iliac artery

Femoral artery

Deep femoral artery

Popliteal artery (behind knee)

Anterior tibial artery

Dorsalis pedis artery

Arteries of the lower limb.

> **Body Language**
>
> The diamond-shaped region behind the knee bordered above by portions of the semimembranosus and semitendinosus (medially) and the biceps femoris (laterally) and below by the two heads of the gastrocnemius muscle, is called the **popliteal fossa**. Within this fossa are the popliteal vein and artery and the tibial and common fibular nerves.

The posterior tibial artery descends in the posterior compartment of the leg in company with the tibial nerve. It gives rise to the fibular artery, which descends on the lateral side of the leg. The fibular artery provides branches that supply the muscles of the lateral compartment.

The posterior tibial artery passes behind the medial malleolus to enter the sole of the foot, where it terminates by dividing into the medial and lateral plantar arteries that supply the muscles and skin of the sole of the foot.

The venous return of blood from the lower limb, like in the upper limb, involves superficial and deep veins and the perforator veins connecting these two systems. These veins contain numerous one-way valves that, through pumping actions of nearby muscles, moves the column of venous blood from the lower limb toward the heart. The superficial veins, called the greater and lesser saphenous veins, begin on the dorsum of the foot from the dorsal venous arch. The greater saphenous vein, the longest vein in the body, begins its climb from the medial side of the foot in front of the medial malleolus. It passes along the medial side of the leg and thigh before emptying into the femoral vein in the upper thigh. The lesser saphenous vein arises from the lateral side of the dorsum of the foot and courses upward on the back of the calf to enter the *popliteal fossa*. Here it empties into the popliteal vein. The popliteal vein is a deep vein. It becomes the femoral vein when it enters the thigh.

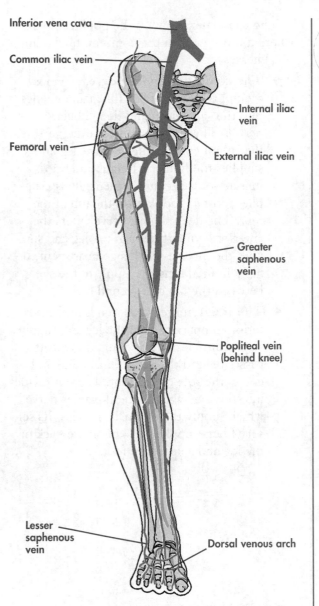

Inferior vena cava

Common iliac vein

Femoral vein

Internal iliac vein

External iliac vein

Greater saphenous vein

Popliteal vein (behind knee)

Lesser saphenous vein

Dorsal venous arch

Veins of the lower limb.

Bodily Malfunctions

It is common for the superficial veins in the lower limb to become enlarged and twisted and appear as bulges through the skin. When this occurs, the condition is known as varicose veins. This happens when the valves within these veins become weakened and allow blood to flow backward and pool. This stretches the veins, producing these unsightly swellings.

Nerve Supply to the Lower Limb

Three nerves supply the musculature and skin of the thigh, leg, and foot:

◆ **Femoral nerve.** Arises from ventral rami of L2, 3, and 4 spinal nerves. This nerve enters the thigh by passing beneath the inguinal ligament. Its fibers supply the anterior compartment muscles and much of the skin on the front of the thigh, leg, and medial side of the foot.

◆ **Obturator nerve.** Also contains fibers from L2, 3, and 4 levels. It supplies the muscles of the medial or adductor compartment and a small area of skin along the upper medial thigh.

◆ **Sciatic nerve.** The largest nerve in the body. It consists of fibers from ventral rami of L4–S3. It leaves the pelvic cavity through the greater sciatic foramen and enters the gluteal region. From here it passes downward into the posterior compartment of the thigh and supplies the muscles in this compartment.

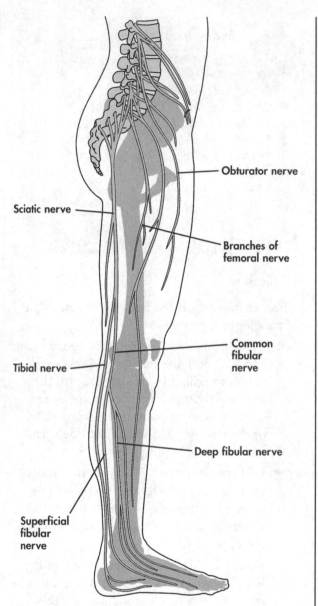

Obturator nerve

Sciatic nerve

Branches of
femoral nerve

Common
fibular
nerve

Tibial nerve

Deep fibular nerve

Superficial
fibular
nerve

Nerves of the lower limb.

The sciatic nerve enters the popliteal fossa, where it terminates in the common fibular and tibial nerves:

♦ **The common fibular nerve.** Wraps around the neck of the fibula and divides into the superficial and deep fibular nerves. The superficial fibular nerve enters the lateral compartment of the leg and supplies the fibularis longus and brevis muscles of this compartment. It also supplies most of the skin on the top of the foot. The deep fibular nerve enters the anterior compartment of the leg and supplies the muscles of this compartment. It has a limited sensory supply to the skin between the big and second toes.

♦ **The tibial nerve.** Descends in the posterior compartment of the leg and supplies all the muscles in this compartment. It passes behind the medial malleolus and enters the sole of the foot. There it divides into the medial and lateral plantar nerves, which supply the intrinsic muscles. Its sensory fibers supply the skin on the back of the leg and sole of the foot.

The Least You Need to Know

◆ The lower limb consists of the hip, thigh, leg, and foot.

◆ The femoral artery provides the blood supply to the lower limb.

◆ The femoral, obturator, and sciatic nerves supply the muscles of the lower limb.

In This Chapter

- ◆ Introduction to the skull and its openings
- ◆ Overview of the five body senses and the anatomy related to them
- ◆ Main contents of the neck

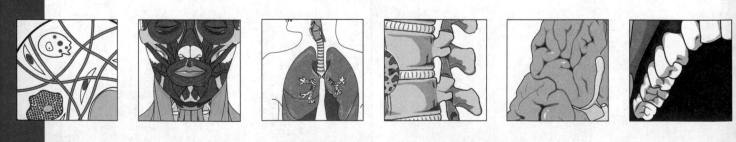

Chapter 21

"On Top of Old Shoulders": The Head and Neck

The expression "big things come in small packages" seems appropriate to describe the sheer volume of anatomic information contained in such a relatively small region of the body. This may explain why there are so many different clinical specialties and subspecialties related to the head and neck.

I don't know if it was by design or by chance, but this last chapter deals with my favorite area of the body to teach. In thinking about how to approach and organize this chapter, I thought about how unique the head was in that it is the headquarters of the body's five senses: sight, hearing, balance, smell, and taste. It seemed logical to highlight the anatomy of the head in the context of the "stories" about these senses. In this regard, I believe the learning process will be easier and more enjoyable and hopefully will stimulate you to learn more on your own.

A Head Case: The Skull and Its Openings

The adult skull consists of 28 bones, including the 6 middle ear ossicles located within the 2 middle ear cavities. Most bones of the skull are connected to one another by fibrous joints called sutures. Synovial joints occur between the mandible (lower jaw bone) and the temporal bone of the skull, as the temporomandibular or jaw joint (TMJ). Synovial joints also occur between the ossicles to permit them to vibrate and enable the sense of hearing (more on that later in the chapter).

The skull can be divided into two functional parts:

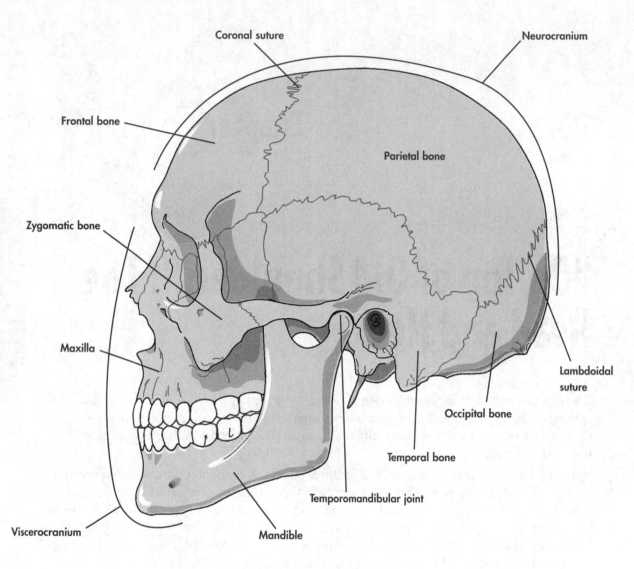

A lateral view of the skull.

◆ **Neurocranium.** This consists of the floor, sides, and roof of the cranium. The latter two together are called the calvaria or "skullcap." The neurocranium contains and protects the brain.

◆ **Viscerocranium.** This is also known as the facial skeleton. It represents the rest of the skull. The viscerocranium contains and protects the upper parts of the digestive and respiratory systems.

Foot Notes

If you think of the skull's neurocranium portion as a rigid, unyielding bony box, it will help you understand why "space-occupying lesions," such as bleeding within the cranium, swelling of the brain following head injuries, growth of tumors, or infections, can put pressure on the brain and alter its function.

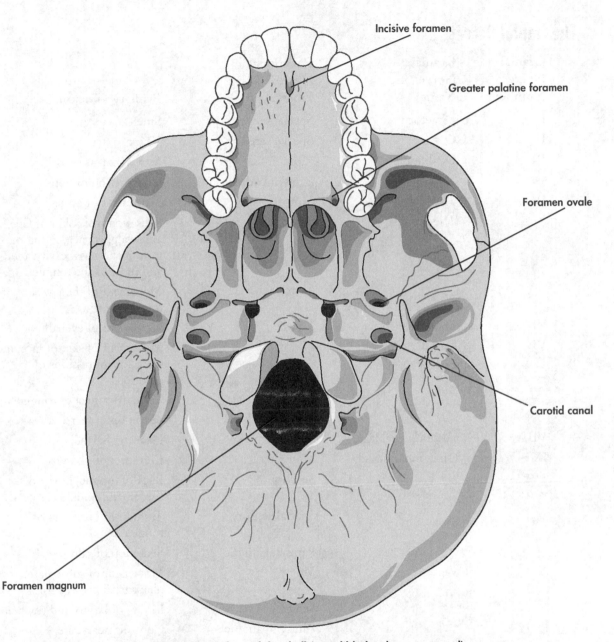

View of the undersurface of the skull (mandible has been removed).

The floor of the neurocranium contains numerous openings and holes called *foramina* (singular, foramen). Cranial nerves from the brain and their branches leave through specific holes to be distributed to the various areas of the head to supply muscle, glands, skin, or mucous membrane. Blood vessels also pass through many of these foramina. The following table lists the names of the cranial nerves, the types of functional fibers each contains, and their principal functions. Additional details regarding these nerves will be given in the following sections.

The Cranial Nerves

Cranial Nerve Number	Cranial Nerve Name	Functional Nerve Fiber Type	Primary Function
I	Olfactory	Special sensory	Smell
II	Optic	Special sensory	Vision
III	Oculomotor	Motor	Movement of eyeball
		Parasympathetic	Constriction of pupil
IV	Trochlear	Motor	Movement of eyeball
V	Trigeminal	Sensory	Sensation from skin of face and scalp, cornea, lining of mouth and nasal cavity, teeth, and most of the tongue
		Motor	Movement of the jaw during chewing
VI	Abducens	Motor	Movement of eyeball
VII	Facial	Motor	Movement of skin of face and scalp for expressions of emotion
		Special sensory	Taste over most of tongue
		Parasympathetic	Secretion of tears, saliva
VIII	Vestibulocochlear	Special sensory	Hearing, balance
IX	Glossopharyngeal	Motor	Elevation of pharynx
		Sensory	Back of tongue, lining of throat, and middle ear cavity
		Special sensory	Taste from back part of tongue
		Parasympathetic	Secretion of saliva
X	Vagus	Motor	Movement of soft palate, pharynx, and vocal cords
		Sensory	Lining of larynx and trachea
		Parasympathetic	Controls heart rate, respiration, and glands and smooth muscle in the walls of most of the digestive organs
XI	Spinal accessory	Motor	Supplies muscles that elevate the shoulders and turn the head to the side
XII	Hypoglossal	Motor	Movement of tongue

The Bone-to-Skin Connection: The Muscles of the Face and Scalp

Because the face is always sticking out there in front, it's one of the most commonly injured regions of the body, from car accidents, sports injuries, or physical violence. The skin of the neck and face is thin and supple. But the skin on the scalp is thick. Beneath the skin, within the superficial fascia, is a rich blood supply and cutaneous nerve branches from the trigeminal nerve (CN V).

Within the superficial fascia of the neck, face, and scalp are unique skeletal muscles, collectively called the muscles of facial expression. These are small, delicate muscles that attach

from bones of the skull into the overlying skin. As their name suggests, they move the skin to create the many expressions of emotion that we recognize as human, such as sadness, concern, surprise, joy—emotions that are "written all over our face." These muscles have both psychological and physiological significance because many of them act as sphincters ("closers") or dilators ("openers") of the openings on the face. Several of these muscles are labeled in the accompanying figures. For example, the *orbicularis oculi* is a large circular-shaped muscle that surrounds the orbit (the eye socket) and the eyelids. Contraction of this muscle closes the eyelids, as in winking or blinking. This causes them to act like windshield wipers to distribute a thin film of tears from the lacrimal (tear) gland over the surface of the eyeball to keep the cornea moist and healthy.

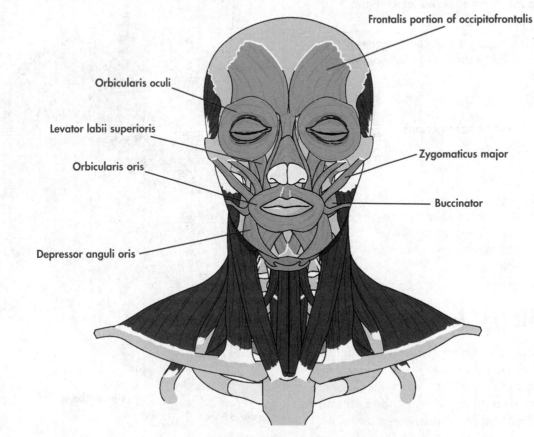

Orbicularis oculi

Levator labii superioris

Orbicularis oris

Depressor anguli oris

Frontalis portion of occipitofrontalis

Zygomaticus major

Buccinator

Muscles of facial expression.

The *orbicularis oris* is a similarly circular-shape muscle surrounding the mouth. It contracts to close the lips. The *buccinator* ("trumpeter") muscle is located within the cheek wall. It compresses the cheek, as when a person forces air out through pursed lips into a wind instrument. Both of these muscles also maintain tension within the lips and cheeks during chewing so that food materials stay in the oral cavity proper and do not fill the space between the lips and the teeth (called the vestibule of the oral cavity—described in a later section).

The scalp also contains a muscle, called the *occipitofrontalis*. This muscle has a pair of bellies, the occipitalis portion and the frontalis portion, that are connected by a thin but tough aponeurosis that covers the top of the calvaria. The occipitofrontalis muscle attaches to the skin over the eyebrows. Contraction of this muscle causes the forehead to wrinkle, as in the expression of surprise.

The nerve supply to all the muscles of facial expression is via branches of the facial nerve (CN VII).

Bodily Malfunctions

Bell's palsy is an inflammatory condition that affects the facial nerve as it passes through the temporal bone. Swelling and compression of this nerve weakens the facial muscles on the affected side, causing them to sag. Inability to shut the eyelids may cause the cornea to dry out if not treated properly.

You Fill Up My Senses

The body has five senses that relate to sight, hearing, balance, taste, and smell. The organs of these senses are delicate structures that need protection from injury. Many of these structures are contained within protective spaces within the head.

I See a Vision: Orbit, Eyeball, and the Sense of Sight

The sense of sight or vision is associated with the eyeball, located within a bony socket called the *orbit*. The orbit is a four-sided pyramid structure. Its base or widest part is located in front and is formed by portions of the frontal, maxillary, and zygomatic bones. Its apex is narrow and is located toward the back of the socket. The orbit contains the optic canal and the superior orbital fissure. The optic canal transmits the optic nerve (CN II) and the ophthalmic artery, a branch of the internal carotid artery, into the orbit. The ophthalmic artery supplies blood to the eyeball, including the retina, and the contents of the orbit.

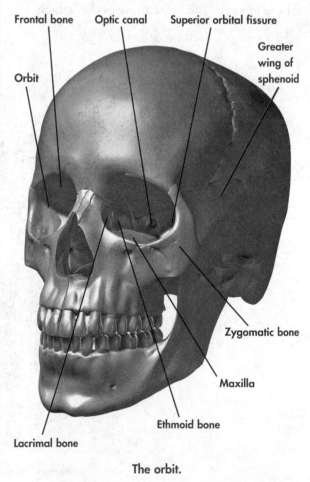

The orbit.

The orbit has four walls:

◆ **Superior wall or roof.** Formed mainly by the frontal bone. The lacrimal (tear-producing) gland is located in the upper outer portion of this wall.

◆ **Lateral (outer) wall.** Formed by the zygomatic bone and greater wing of the sphenoid.

◆ **Inferior wall or floor.** Formed by the maxillary bone. The bone of the floor of the orbit is thin and vulnerable to fracture following a blow to the orbit.

◆ **Medial (inner) wall.** Formed mainly by the ethmoid and lacrimal bones. This wall is paper thin. Within the ethmoid bone at this site are several air-filled bony cells that are part of the paranasal sinus system in the head (discussed in a later section).

For a detailed look at these walls, see the first page of the color insert section.

The eyeball is surrounded and supported by considerable fat and moves by contraction of the *extraocular muscles*. These are slender strap-like skeletal muscles. They consist of four recti ("straight") muscles: medial rectus, lateral rectus, superior rectus, and inferior rectus; and two oblique muscles: superior oblique and inferior oblique. These muscles attach at one end to bone near the apex of the orbit (except the inferior oblique, which attaches to bone of the floor near the front of the orbit), and at the other end to the sclera or "whites" of the eyeball. These muscles act in a coordinated manner to move the eyeball up and down and in and out.

Bodily Malfunctions

The thin medial and inferior walls of the orbit are vulnerable to fracture from blows to the eyeball. Such fractures are called "blow out" fractures of the orbit. A blow out fracture of the floor may cause contents of the orbit to enter the maxillary sinus located directly beneath the orbital floor. Infections of the maxillary sinus or ethmoid air cells may erode through the bone and involve the orbit.

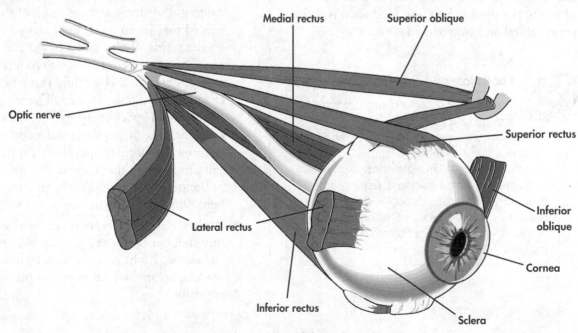

The extraocular muscles of the right eye.

One additional muscle of the orbit, the levator palpebrae superioris (palpebra means "eyelid"), does not attach to the eyeball. Instead, it attaches to the upper eyelid. Decoding its $50 name reveals its function, to elevate the upper eyelid. It is the antagonist to the orbicularis oculi muscle, which closes the eyelid.

Three different nerves supply the extraocular muscles as follows:

- **Oculomotor nerve (CN III).** Supplies the superior rectus, medial rectus, inferior rectus, inferior oblique, and levator palpebrae superioris.

- **Trochlear nerve (CN IV).** Supplies just one muscle, the superior oblique.

- **Abducens nerve (CN VI).** Supplies just one muscle, the lateral rectus.

These nerves enter the orbit from inside the neurocranium by passing through the superior orbital fissure. Other nerves, branches of the ophthalmic division (CN V_1) of the trigeminal nerve (CN V), relay sensory information from the eyeball (via the nasociliary nerve), the lacrimal gland (via the lacrimal nerve), and the skin of the upper eyelid, forehead, and scalp (via the supraorbital and supratrochlear nerves).

Foot Notes

The corneal reflex is a protective reflex that guards the cornea from injury. It involves a sensory limb provided by the nasociliary nerve to the cornea, and an efferent (motor) limb provided by the facial nerve to the orbicularis oculi muscle. Injury to either leaves the cornea vulnerable because a person won't automatically close her eyes when she sees something heading for her.

The optic nerve (CN II) is actually an outgrowth of the brain during development and is specialized for the sense of vision. It starts at the retina, which contains the photoreceptors called the *rods and cones*. Light rays entering the eyeball through the pupil are focused onto the retina. Stimulation of these cells causes photochemical reactions within the retina that are converted to electrical impulses. These are conveyed along the optic nerve to the brain for processing and interpretation.

The eyeball consists of three layers:

- **Outer fibrous layer.** Consists of the sclera or the white part of the eyeball. It is opaque and makes up the posterior five sixths of the eyeball. The cornea is the anterior one sixth. It is transparent, lacks a blood supply, and is richly supplied by sensory nerve fibers—a fact well confirmed by anyone who has experienced a scratched cornea!

- **Middle layer.** A vascular and pigmented layer. It contains the choroid, ciliary body, and the iris. The ciliary body contains a smooth muscle (the ciliary muscle) that controls the shape and thickness of the lens so that it can focus light rays properly on the retina. It also contains the ciliary process that continuously produces aqueous humor, a clear fluid that fills the chamber behind the cornea. The iris is the "colored" part of the eye. It contains a dilator (dilator pupillae) and a constrictor muscle (sphincter pupillae). Both are smooth muscles that control the diameter of the pupil and regulate the amount of light entering the eye.

- **Retina.** A complex, many-layered structure that contains the light-sensitive rods and cones. Light rays are focused onto the macula, an area which produces the sharpest vision.

There are four refractive media that bend the light rays entering the eyeball:

◆ **Cornea.** The principal "light bender."

◆ **Aqueous humor.** A clear watery solution that is continuously produced by the ciliary processes and reabsorbed into the venous system. Overproduction or reduced reabsorption causes an increase in pressure within the eye, which decreases the blood supply to the retina. This is known as glaucoma, and can silently lead to blindness.

◆ **Lens.** Located behind the iris. It is a transparent, biconvex elastic structure. It changes shape to focus the light rays onto the retina. Loss of its transparency is known as a cataract.

◆ **Vitreous body.** Located behind the lens. It consists of a transparent jelly-like fluid, the vitreous humor, which fills the vitreous chamber between the lens and retina and gives the eyeball its shape. This substance does not turn over, meaning that loss of this substance through a tear in the eyeball leads to blindness in that eye.

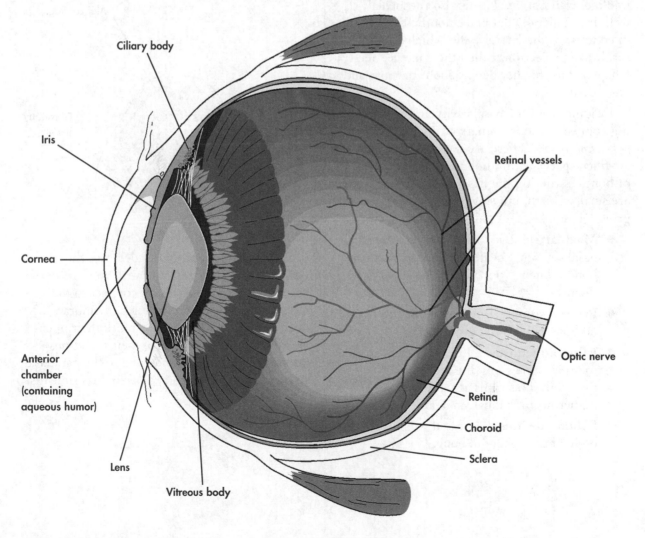

The parts of the eyeball.

I Didn't Odor This!: Nose, Nasal Cavity, and Sense of Smell

When you look at a dry skull, the "nose" is missing! Where did it go? The portion of the nose that appears on a person's face is made of cartilage and dense fibro-fatty tissue, which is tightly covered by skin and lined by mucous membrane. These tissues are attached to an opening on the front of the skull, called the *anterior nasal aperture*, which is made by the maxillary and nasal bones. The nasal cavity is divided into left and right nasal cavities by the midline *nasal septum*. This is also the medial wall. It is generally flat and smooth. This is in contrast to the lateral walls, which contain scroll-like projections called the conchae. See Chapter 9 for further details about the nose and nasal cavity.

Leading from the nasal cavity and maintaining connection to it via narrow openings are the *paranasal sinuses*. These are membrane-lined, air-filled spaces within hollowed-out portions of bones bordering the nasal cavity. The sinuses are small at birth, but slowly grow as the person grows. There are four pairs of paranasal sinuses:

- ◆ **Maxillary sinus.** The largest of the paranasal sinuses. Located within the maxillary bones. Their "roof" forms the "floor" of the orbit.
- ◆ **Frontal sinus.** Located within the frontal bone behind the eyebrows.
- ◆ **Ethmoid sinuses.** Located behind the thin bone along the medial walls of the orbit. This sinus actually consists of numerous individual small air cells.
- ◆ **Sphenoid sinus.** Located within the body of the sphenoid bone.

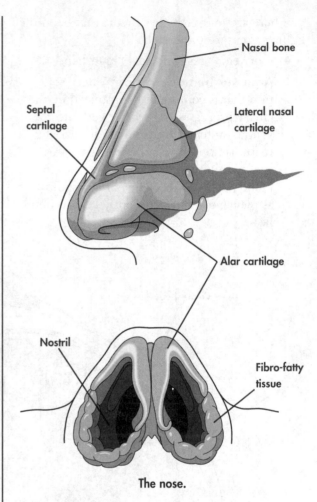

The nose.

Presumably, these sinuses decrease the weight of the skull and help with voice resonance. Because these sinuses connect to the nasal cavity, swelling of the nasal membranes due to colds or allergies may block their openings, preventing drainage of their secretions into the nasal cavity. When one of the paranasal sinuses becomes infected, a person has sinusitis.

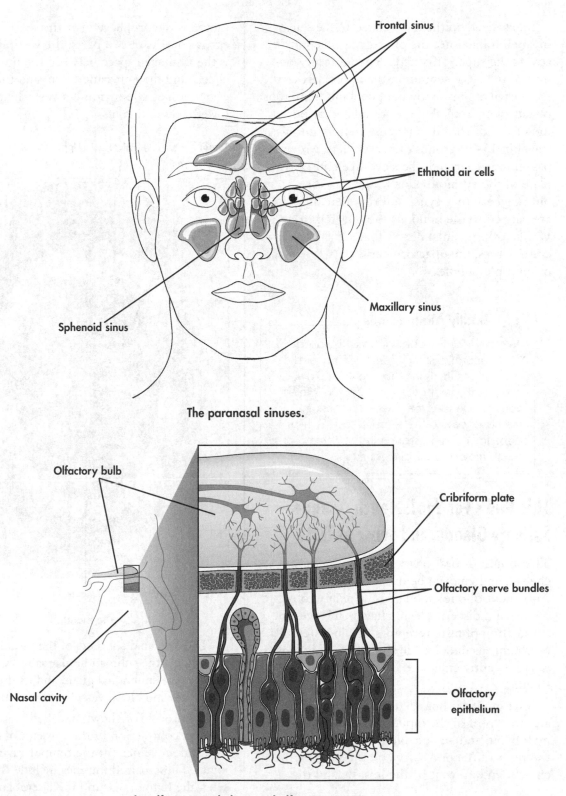

The paranasal sinuses.

The olfactory epithelium and olfactory nerve.

Specialized epithelium related to the sense of smell (olfaction), the olfactory epithelium, covers the superior conchae and adjacent portion of the nasal septum in the upper part of the nasal cavity. This epithelium contains numerous olfactory neurons that are stimulated by odor molecules dissolved in mucus. Sensory impulses travel in 15–20 groups of nerve fibers, which pass upward through the sieve-like cribriform plate of the ethmoid bone to the olfactory bulbs, where they synapse. From here, impulses are carried to the brain via the olfactory tracts. Unlike other cranial nerves that are single, discrete nerves, the olfactory nerve (CN I) is made of multiple bundles.

Bodily Malfunctions

Loss of the ability to smell is called anosmia. It may result from a head injury involving the cribriform plate, complications of an upper respiratory infection, or due to some disease process in the nasal cavity. The sense of taste, which depends a great deal on the ability to smell, may also be diminished.

This Bud's For You!: Mouth, Tongue, Salivary Glands, and Sense of Taste

The mouth or oral cavity is where the food is chewed and enjoyed before it is swallowed. The enjoyment of one's food and the ability to recognize and discern specific flavors involves the stimulation of taste receptors within taste buds by chemicals dissolved in the saliva. This sense is significantly enhanced by an intact sense of smell.

The mouth is bounded above by the palate (the "roof of the mouth"), below by the mylohyoid and geniohyoid muscles that form the floor of the oral cavity, on the sides by the cheeks, and in front by the lips. Behind the

mouth is the oropharynx or throat. The oral cavity consists of two parts: the vestibule, which is the U-shaped space between the lips and cheeks and the gums and teeth; and the oral cavity proper, the region between the upper and lower teeth and gums.

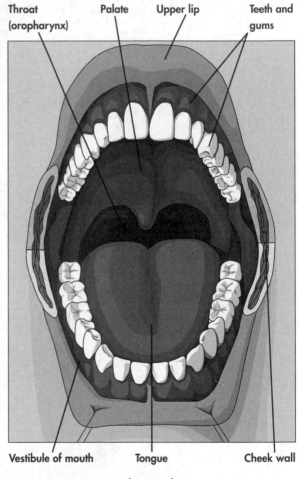

The mouth.

The principal contents of the mouth include the tongue, the submandibular salivary gland and its duct, the sublingual gland and its ducts, and the nerves and blood vessels of the oral cavity.

The process of chewing, called mastication, involves contraction of the muscles of mastication and movement at the temporomandibular joints. These paired muscles include the masseter, the temporalis, and the lateral and medial pterygoid muscles. If you clench your jaw, you can feel the stiffened masseter muscles on the sides of the face in front of the ears. You can also feel the *temporalis* muscles located on the sides of the head in the temple region. These arc large, strong muscles that elevate or close the mandible during chewing. The *pterygoid* muscles are located deep to the ramus of the mandible and can't be felt. They protrude the jaw during the early stage of jaw opening and move the jaw from side to side in a grinding motion during chewing. These muscles receive their nerve supply from the mandibular division (CN V₃) of the trigeminal nerve (CN V).

The tongue also participates in the process of chewing and breaking down the food materials to a consistency safe for swallowing. In fact, its function is essential to this process. It moves around the food we chew so that it gets crushed by the teeth. Hence, the tongue is mobile. It is a skeletal muscle consisting of many *extrinsic* and *intrinsic* muscles. The extrinsic muscles attach to different bones before connecting to the tongue. They include the following:

Body Language

The **extrinsic muscles** of the tongue alter the position of the tongue. The **intrinsic muscles** of the tongue begin and end in the tongue and do not attach to bone. They change the shape of the tongue.

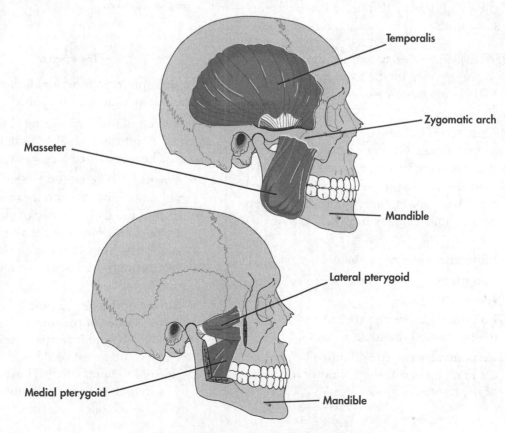

The muscles of chewing (mastication).

◆ **Genioglossus.** "Genio" refers to chin. This is a large, fan-shaped muscle that attaches in front to the inside of the mandible and fans out into the tongue. It is the principal muscle of the tongue and helps protrude it. It is the muscle you use when you stick out your tongue or move it from side to side.

◆ **Hyoglossus.** A muscle that attaches to the sides of the tongue.

◆ **Styloglossus.** A muscle that attaches to the back and sides of the tongue. It helps to elevate and pull the tongue back into the mouth.

◆ **Palatoglossus.** It attaches from the sides of the soft palate to the sides of the tongue. It helps to raise the tongue toward the soft palate and seal off the mouth from the throat during chewing so that you can chew and breathe through the nose at the same time.

All of these muscles, except for the palatoglossus, are supplied by the hypoglossal nerve (CN XII). The palatoglossus muscle is innervated by the vagus nerve (CN X).

The surface of the tongue is covered by mucous membrane. On the part of the tongue located in the mouth (the oral part), this membrane contains numerous papillae that give the tongue a velvety appearance. There are a variety of different papillae, many of which contain taste buds:

◆ **Filiform.** The most abundant type.

◆ **Fungiform.** Are "mushroom-like" in appearance.

◆ **Foliate.** Located on the sides of the tongue toward the back

◆ **Circumvallate.** Are "donut-like"; these are lined up in a V shape toward the back of the tongue.

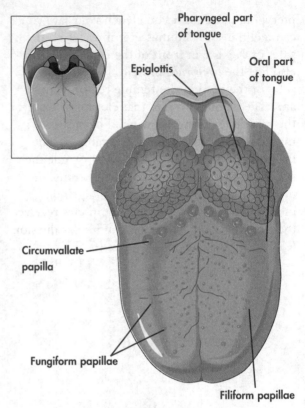

The tongue.

In addition to its function during chewing, the tongue is important for speech.

Softening of the food during chewing is aided by an infusion of saliva from the salivary glands. There are three major salivary glands:

◆ **Parotid.** The largest of the three, although it produces only about 25 percent of the daily volume of saliva. The parotid duct opens at the vestibule of the oral cavity after crossing the face and passing through the buccinator muscle and the cheek wall.

◆ **Submandibular.** Located below the lower border of the mandible, where it can be felt, and it wraps around the back edge of the mylohyoid bone to enter the floor of the oral cavity. Here, it gives rise to a long duct called the submandibular duct, which opens on either side of the

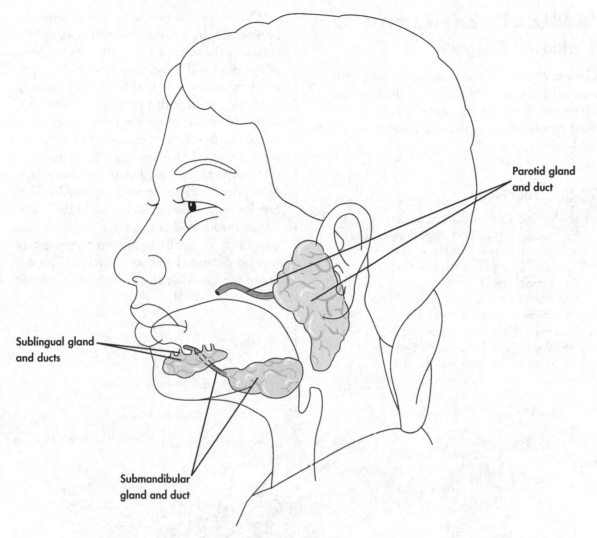

Parotid gland
and duct

Sublingual gland
and ducts

Submandibular
gland and duct

The salivary glands.

frenulum, a tissue fold that connects the undersurface of the tongue to the floor of the mouth. This gland produces about 70 percent of the daily volume of saliva.

◆ **Sublingual.** Located forward on the floor of the mouth next to the tip of the tongue. It has about a dozen small ducts that open on the mucous membrane. It produces the remaining 5 percent of the saliva.

Bodily Malfunctions

Occasionally, the salivary glands produce stones, called sialoliths. These are the size of grains of sand, but can block the narrow ducts of these glands and produce pain and swelling. A common site of blockage is at the opening of the submandibular duct on the floor of the mouth beneath the tongue. The timing of these bouts of painful swelling coincides with mealtimes, when salivary flow is greatest.

I'm All Ears: The Ear and Sense of Hearing and Balance

The ear consists of three anatomic parts: the external ear, the middle ear, and the inner ear. The external and middle ear regions are air-filled spaces, whereas the inner ear region is a fluid-filled space.

Foot Notes

There are two kinds of hearing loss, conductive and sensorineural. The conductive type involves the sound-conducting air-filled spaces of the external and middle ears. Examples include small objects or excess wax lodged in the external auditory meatus or fluid buildup within the middle ear. Sensorineural hearing loss involves injury to the cochlea (the organ of hearing) or the cochlear nerve.

The external ear consists of two parts: the auricle and the external auditory meatus. The auricle is the structure that attaches to the side of the head and is continuous with the external auditory meatus or ear canal. It is composed of elastic cartilage that is tightly covered by skin. The lobule of the ear (where earrings are hung) is a fat-filled appendage that hangs down from the ear. The external auditory meatus is an S-shaped canal that has an outer cartilaginous portion and an inner bony portion. This canal is only about an inch long in the adult. At the medial end of this canal is the tympanic membrane or eardrum. Special ceruminous or wax-producing glands are contained in the lining of the canal for entrapment of dust or debris entering the canal.

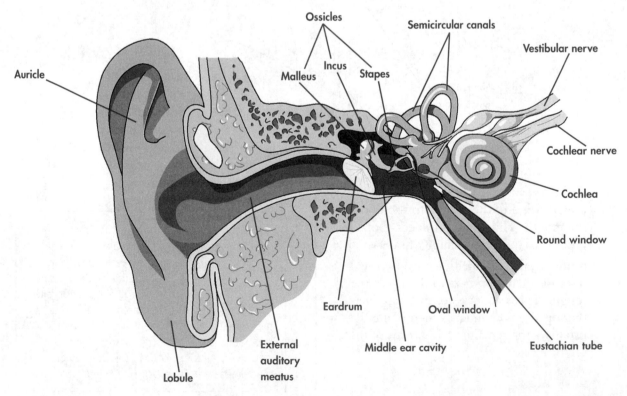

The ear.

The eardrum forms the outer wall of a box-like middle ear cavity. The eardrum is a thin, pearly gray colored translucent membrane that is about one centimeter wide. Within the middle ear cavity is a chain of tiny bones, called the *auditory ossicles*. There are three of these in each middle ear:

- **Malleus.** Looks like a hammer. The outermost bone of the chain, it attaches to the inner surface of the eardrum.
- **Incus.** Looks like an anvil. The middle bone, it attaches to the innermost bone, the stapes.
- **Stapes.** Looks like a stirrup. At the end of the stapes is its footplate, which attaches to the margin of the oval window of the inner ear.

These bones are suspended from the walls of the middle ear by equally tiny ligaments. They articulate with each other through synovial joints.

There are two muscles within the middle ear cavity:

- **Tensor tympani.** Attaches to the malleus bone and pulls it, along with the eardrum, medially to stiffen the membrane and reduce its vibration to protect against loud noises.
- **Stapedius.** The smallest muscle in the body, it attaches to the stapes and draws the footplate back to reduce vibration across the oval window.

Hearing involves the reception of sound waves by the auricle and their passage down the auditory meatus to the eardrum, causing it to vibrate. Vibration of the eardrum sets the ossicles into motion. The stapes, attached to membrane at the oval window of the inner ear, acts in a piston-like fashion to transmit its mechanical energy to the fluid in the inner ear. Specialized cells of hearing in the cochlea become stimulated. They convert the mechanical energy to electrical impulses. These impulses are sent to the central nervous system via the cochlear nerve for processing and interpretation.

Bodily Malfunctions

Abnormal growth of bone around the footplate of the stapes causes it to become fixed and unable to move at the oval window. This condition is known as otosclerosis. It's a common cause of conductive hearing loss.

The middle ear cavity connects to the throat via the Eustachian or auditory tube. This tube is normally closed where it connects to the throat and opens temporarily when a person swallows. This allows air to rush into the middle ear and equalize pressure on both sides of the eardrum. We all have experienced situations while flying or traveling in elevations when our ears suddenly feel "full" and uncomfortable. This is due to unequal pressures across the eardrum, which is alleviated by frequent swallowing or yawning to open the auditory tube.

Bodily Malfunctions

The auditory tubes are shorter and more horizontal in young children. Upper respiratory infections at this age frequently lead to infections of the middle ear, called otitis media. During these infections, fluids are produced that may cause the eardrum to bulge outward. In addition to causing pain, they may temporarily interfere with hearing.

The inner ear consists of two structures with different functions:

◆ Cochlea

◆ Vestibular apparatus

These structures are parts of the *bony labyrinth*, a complex system of hollowed-out canals and spaces within the temporal bone. The bony labyrinth is filled with fluid, called perilymph. Within this fluid are suspended membrane structures filled with endolymph fluid. These structures make up the *membranous labyrinth*. These include the cochlear duct, containing the sense organs for hearing, and the utricle and saccule. They also include the three semicircular canals, which contain sense organs for balance and motion. Movements of the fluid within these spaces stimulate sensory cells within the membranes to produce electrical signals. These signals are carried to the central nervous system along the cochlear nerve (for hearing) and the vestibular nerve (for balance) branches of the vestibulocochlear nerve (CN VIII).

The Neck: The Great Connector

Several structures that begin in the head pass through the neck on their way to the thorax. These include the pharynx and esophagus, the larynx and trachea, the internal jugular veins returning blood from the brain and portions of the head and neck, and the vagus nerves. Other structures, such as the sympathetic trunks and the right and left common carotid arteries, ascend from the thorax into the neck. The common carotid arteries divide into external and internal carotid arteries near the upper border of the thyroid cartilage of the larynx. Branches of the external carotid arteries provide much of the blood supply to the neck, face,

scalp, and head while branches of the internal carotid arteries supply blood to the brain, orbit, and a portion of the scalp.

Cervical Triangles

The neck is classically divided into anterior and posterior triangles by the sternocleidomastoid muscles. These muscles pass diagonally downward in the neck from a point on the skull behind the ear to the clavicle and sternum. The posterior triangle is bounded by the sternocleidomastoid muscle, the trapezius muscle of the shoulder, and the clavicle. The anterior triangle is bounded by the sternocleidomastoid, the lower border of the mandible, and a vertical line connecting the chin to the sternum. There are also several smaller named triangles. These various triangles contain specific structures, and thus serve to guide surgical approaches to injured or diseased contents.

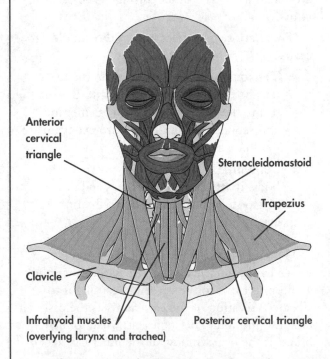

The anterior and posterior cervical triangles.

Cervical Viscera (Neck Organs)

In the midline of the neck are the cervical viscera. In front is the larynx, or the voice box. It was described in detail in Chapter 9. In the neck, the thyroid cartilage and the arch of the cricoid cartilage can be felt. At the midpoint along the front edge of the thyroid cartilage is the approximate location of the vocal cords. Connecting these two cartilages is a thick membrane called the *cricothyroid membrane* that also can be felt along the midline.

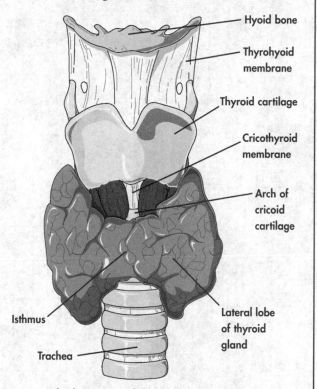

The larynx, trachea, and thyroid gland.

Labels:
- Hyoid bone
- Thyrohyoid membrane
- Thyroid cartilage
- Cricothyroid membrane
- Arch of cricoid cartilage
- Lateral lobe of thyroid gland
- Isthmus
- Trachea

Foot Notes

Occasionally, in an extreme emergency, when a person is choking on a piece of food that has gotten stuck at the vocal cords, an emergency airway can be made by an incision through the cricothyroid membrane. This procedure is known as a cricothyrotomy.

The trachea continues downward into the thorax from its attachment to the larynx.

Resting on the larynx is the *thyroid gland*. It is an endocrine gland that secretes thyroxine for regulating metabolic activities in the body and calcitonin for regulation of blood calcium levels. It consists of a pair of lateral lobes that extend downward along the trachea. The lobes connect in front of the upper trachea by an isthmus. Located on the posterior surfaces of the lobes are four small, pea-sized *parathyroid glands*. They are also endocrine glands that produce parathyroid hormone. This hormone acts on bone cells for calcium and phosphorus metabolism.

Behind the larynx is the lower end of the pharynx, the laryngopharynx. From here the esophagus begins and also passes downward into the thorax.

The Least You Need to Know

- The skull consists of the neurocranium, a rigid portion of the skull that protects the brain, and the viscerocranium, which contains the upper parts of the respiratory and digestive systems.

- The floor of the skull contains numerous holes through which pass the cranial nerves and their branches and blood vessels.

- The head contains the organs for the body's five senses: vision, smell, taste, hearing, and balance.

- The neck contains several cervical viscera, the carotid arteries, internal jugular veins, and vagus nerves.

Appendix A

Glossary

abdomen Also known as the belly, this is the region of the body between the thorax and the pelvis.

abduction To move away from the midline of the body in the coronal plane.

accessory organs The accessory organs of the digestive system include the liver, the gallbladder, and the pancreas. They help digest and metabolize the nutrients from the digestive tract.

acetabulum The "cup" or socket that holds the head of the femur at the hip joint.

adduction To move toward the midline of the body in the coronal plane.

adipocytes Also known as fat cells, these are specialized connective tissue cells that store lipids (fatty acids). Their cytoplasm is mostly a single droplet of fat.

afferent fibers Neurons that send impulses toward the spinal cord and brain.

alveoli Air sacs that are the structural and functional unit of respiration. There are approximately 300 million of these air spaces in each lung.

anatomical position The person is standing erect, looking straight ahead, with arms at the sides and palms facing forward. The legs are together, with toes directed forward. Structures of the body are described in relationship to this position.

anatomical sphincter A valve that is made of a thick ring of smooth muscle. Passage of material through it is controlled by autonomic nerves.

anterior A structure that is anterior (or ventral) is closer to the front of the body than a structure that is posterior (or dorsal), which lies closer to the back of the body.

anterior compartment   Contains muscles located on the front side of the bone(s) in the upper and lower limbs.

aorta   One of the great vessels of the heart. The aorta receives oxygen-rich blood from the left ventricle and sends it to the body by way of its systemic branches.

apocrine sweat glands   Limited in their distribution—they are located in the skin of the armpit, the pubic region, and the pigmented area of the breasts. After puberty, apocrine glands begin secreting a slightly viscous fluid that develops an odor following contact with bacteria on the surface of the body.

aponeuroses   Thin tendonlike membranes that connect muscles to one another or to bones.

appendicular skeleton   The part of the skeletal system consisting of the appendages.

aqueous humor   A clear, watery solution that is continuously produced by the ciliary processes in the eye and reabsorbed into the venous system.

arachnoid mater   A delicate, fibrous membrane that is loosely attached to the inner surface of the dura mater in the meninges.

areola   The pigmented area around the nipple.

arterial blood   Oxygen-rich blood that is leaving the heart for the body.

articular processes   Bony parts of vertebrae, which have upper (superior) and lower (inferior) versions, that stick out upward or downward from each pedicle.

articulation   Another word for joint, where two bones come together.

ascending pathways   Nerve highways that send sensory information up to the brain from the body.

association areas   Parts of the brain used for more complex mental functions such as recognizing objects.

atlas   The top vertebra in the neck (also called C1). It is oval and has no body.

atria   The upper chambers of the heart, left and right.

auditory ossicles   A tiny chain of three bones within the middle ear cavity: malleus, incus, and stapes.

autonomic nervous system   A functional division of the peripheral nervous system that supplies nerves to the internal organs. "Autonomic" means, roughly, "not under voluntary control."

axial skeleton   The axis of the human skeletal system, consisting of the skull, spine, ribs, and breastbone.

axilla   Armpit.

axillary artery   The primary artery to the upper limb. It's a continuation of the subclavian artery and gives rise to several smaller arteries, such as the two circumflex humeral arteries.

axis   The second neck vertebra (C2). It helps support the atlas and enables you to turn your head from side to side.

axon   A single nerve process that carries impulses away from the cell body. An axon is a long process (some axons may be up to a few feet in length in tall people!) that ends by dividing into several terminal branches.

basal nuclei   The correct term for the collection of nerve cell bodies inside the cerebral hemispheres.

biceps brachii   This "two-headed" muscle in the anterior compartment flexes and supinates the forearm.

bladder   A temporary storage reservoir for urine until the person decides to empty it.

bony labyrinth   A complex system of hollowed-out canals and spaces within the temporal bone. The bony labyrinth is filled with fluid, called perilymph.

brachial plexus   A network of nerves formed by the spinal nerves from the C5, C6, C7, C8, and T1 levels of the spinal cord. From this plexus arise the nerves that supply the muscles and skin of the upper limb.

brachialis   A muscle in the anterior compartment that flexes the forearm. The brachialis attaches to the humerus on one end and the ulna on the other.

brain stem   Located in the center of the brain beneath the cerebral hemispheres. It's a more primitive part of the brain that controls breathing, blood pressure, and heart rate. In addition, all sensory and motor information passing to and from the cerebral cortex goes through the brain stem.

bronchi   The two divisions of the trachea that enter the lungs and branch into ever smaller bronchi.

bronchioles   The smallest air passages in the lungs.

bulbourethral glands   Also known as *Cowper's glands,* two pea-size glands located near the root of the penis. They secrete an alkaline substance that lubricates the urethra and neutralizes the acid from urine.

canaliculi   The slender, fluid-filled channels that connect osteocytes to one another and to the osteoblasts.

cancellous bone   See *trabecular bone.*

capillaries   Located in most tissue spaces of the body and connect small arteries to small veins. They are the body's narrowest and thinnest blood vessels, essentially consisting of tubes of endothelial cells.

cardiac muscle   A muscle type found only in the heart. It is the principal tissue forming the walls of the heart's chambers. It's an involuntary muscle type.

cardiac veins   A system of veins in the heart that return venous blood to the right atrium.

carpal tunnel   An enclosed space at the wrist through which the tendons of the flexor muscles to the fingers and thumb enter the palm of the hand. This fibrous-bony tunnel is formed by a tough sheet of connective tissue, the flexor retinaculum, which attaches to and arches across some of the carpal bones.

carpals   The eight irregularly shaped bones of the wrist.

cauda equina   The long "horse's tail" of the lowermost nerve roots in the spinal cord.

cell body   The part of the nerve cell that contains the nucleus.

central nervous system   The brain and the spinal cord. The CNS receives, decodes, processes, and responds to information it receives from the body, by means of the *peripheral nervous system (PNS).*

cerebellum   The part of the brain located beneath the occipital lobes of the cerebrum. It maintains balance and posture. It also controls the strength and duration of muscle contraction so that an action is carried out smoothly and with precision.

cerebrospinal fluid   The clear liquid that cushions and protects the brain and spinal cord.

cerebrum   The largest part of the brain, where logic and reasoning occur.

chondroblasts   Cells that produce cartilage matrix (framework) and create a model of the bone, which is later mineralized.

chyle   Lymph in the digestive system that contains fatty acids and appears milky.

cilia Membrane-covered, moving structures most common in the lining of the nasal cavities and the upper respiratory system. Their wave-like motion sweeps away inhaled dust particles.

circumduction The universal type of free movement at ball-and-socket joints, like the shoulder and hip.

clavicle Collarbone.

coccyx The "tail end" of the spine. It consists of three to five (usually four) small vertebrae that have lost their resemblance as vertebrae. Rather, they appear as segments of solid bone connected to one another by small amounts of fibrocartilaginous tissue.

collagen The most abundant protein in the body. It is found in high concentrations in structures that require tensile strength and rigidity, such as bones, teeth, tendons and ligaments, and the dermis of the skin.

connective tissue The "glue" that provide the structural and nutritional support for cells, tissues, and organs of the body.

conus medullaris The bottom tip of the spinal cord.

cornea The principal "light-bender" of the eye. It is transparent and part of the outer layer of the eye.

coronal or frontal plane Passes through the body at a right angle to the median plane. It divides the body into front (*anterior*) and back (*posterior*) parts.

coronary arteries The two main arteries that supply the heart with arterial blood.

cortical bone The dense outer shell on all bones.

costal A Latin word meaning "of or near the ribs."

costal (rib) demifacets A set of small, shallow depressions along each upper and lower edge of the thoracic vertebrae, for attachment of the heads of ribs.

cranial nerve One of the 12 pairs of cranial nerves that come off the bottom of the brain at irregular intervals.

cranial nerves Peripheral nerves that leave the brainstem and floor of the cranium to enter the head and neck.

craniosacral outflow Describes the parasympathetic division of the peripheral nervous system by reference to the location of the cell bodies of its preganglionic neurons in the brain stem and the S2, S3, and S4 segments of the spinal cord.

deep Farther from the surface of the body.

deltoid A very important and powerful muscle of the shoulder joint, which abducts the arm.

dendrites Numerous slender cell surface projections from the cell body. They receive electrical signals or impulses from other nerve cells and relay these impulses toward the cell body.

dense connective tissue Differs from loose connective tissue by having fewer cells and more abundant thick bundles of collagen fibers. Such tissue is strong and well suited to resist stress and forces applied to it.

dermis The layer of skin under the epidermis. It's made of dense, irregular connective tissue that contains collagen and elastic fibers.

descending pathways Nerve highways that send movement instructions from the brain to the rest of the body.

detrusor The smooth muscle wall of the bladder.

diaphysis The shaft area of a long bone.

diarthrodial joints Another term for synovial joints.

diastole A period of relaxation in the heart.

diencephalon The part of the brain located between the cerebrum and brain stem. The two main parts of it are the *thalamus* and *hypothalamus*.

dislocation Injury that occurs when the opposing joint surfaces are no longer in contact with one another.

distal The opposite of proximal: the item farthest from the center or the point of attachment.

dorsal See *posterior*.

dorsal horns The part of the gray matter of the spinal cord that receives central processes of sensory neurons.

dorsal rami Branches of spinal nerves that provide nerve stimulation to the deep-back group of muscles. (The singular form of the word is ramus.)

dorsiflexion Movement of the toes and foot toward the leg, like when someone stands on their heels.

ductus deferens (or vas deferens) During sexual arousal, the spermatozoa enter and are transported along the *ductus* or *vas deferens*. The ductus has a narrow channel and a thick muscular wall. This wall strongly propels the cells along the duct and toward the urethra.

duodenum The first and shortest part of the small intestine, where digestion continues with the help of bile.

dura mater The outermost covering of the meninges.

dural sac The tissue surrounding the spinal cord and spinal nerve roots.

eccrine sweat glands Widely distributed in the skin. They produce a clear, watery secretion that contains dissolved salts and other substances.

efferent fibers Motor axons that send information away from the brain and spinal cord.

elastin fibers Elastic fibers, produced by the protein elastin, are the rubber bands of connective tissue.

endocardium The internal layer of the heart's walls.

endochondral ossification The bone-development process in which *chondroblasts* secrete cartilage matrix and produce a model of the future bone. Subsequently, the shaft region of the future bone is mineralized and invaded by blood vessels. Here, new osteogenic cells break down the cartilage matrix and replace it with bone.

enzymes Specialized proteins that can break down complex molecules (such as proteins, fats, and carbohydrates) into simpler compounds.

epicardium The outer wall of the heart.

epidermis The topmost, and thinnest layer of the skin. It has five layers, or strata, within it.

epididymis A highly coiled tube located along the back surface of the testis, used for storing and maturation of sperm cells.

epiphyses The ends of a bone.

epithelial tissue Tissue that lines the external surface of the body, the body's internal cavities and hollow organs, ducts, and blood vessels.

erector spinae A particularly large and powerful group of deep back muscles in the lower back that help maintain your posture and balance.

eversion When the sole of the foot is moved outward, like when someone tries to stand on the insides of their feet.

extension Straightening a body part or increasing the angle at a joint.

extrinsic hand muscles Muscles that reach the hand via long tendons of muscles located in the forearm. These muscles provide power for a firm grip on objects.

extrinsic muscles of the tongue Muscles that alter the position of the tongue.

false pelvis The widened portion of the pelvis above the pelvic brim.

femur The single bone of the thigh.

fibroblasts   The principal cell type found in connective tissue. They produce and secrete collagen, the most common protein in the body, as well as other noncollagenous proteins.

fibrous cartilage   A mixture of tough, white, fibrous tissue and cartilage cells. It is found in places like between your vertebrae.

fibula   One of two bones of the leg.

fissures   Grooves that divide the lungs into lobes.

flexion   Bending a body part or decreasing the angle at a joint.

foramen magnum   A large opening at the base of the skull where the spinal cord begins.

fossae   Shallow surface depressions on a bone where muscles attach.

gallbladder   A pear-shaped structure that is attached to the undersurface of the liver and stores bile.

ganglion   A collection of nerve cell bodies located outside the central nervous system.

gastric or **gastro-**   Refers to anything in or near the stomach.

gastrointestinal (or GI) tract   Includes the esophagus, stomach, small intestine, and large intestine.

gluteus maximus   The largest muscle of the body, located in the hip/buttocks region.

ground substance   A colorless, gel-like substance that consists mainly of water and non-collagenous proteins. The cells and fibers of connective tissue are embedded in it.

hair follicles   A part of the skin that grows hair by packing old skin cells together and pushing them out toward the surface.

hepatic or **hepato-**   Refers to anything of or related to the liver.

hilum   Root of the lung.

horizontal or transverse plane   Passes through the body at right angles to the median and coronal planes and divides the body into upper (*superior*) and lower (*inferior*) parts.

hormones   Chemicals produced by endocrine cells. They are carried in the bloodstream and regulate the activities of other distant cells.

humerus   The bone of the arm.

hyaline cartilage   Cartilage that is gristly and firm, but pliable. It is commonly found on growth plates. In life, hyaline cartilage is bluish-white in color.

hyaluronic acid   What makes the synovial fluid stringy. It also coats each cell in the cartilage and makes the matrix more resilient.

hypodermis   Connects the skin to the underlying muscles of the body. It is made of loose connective tissue and variable numbers of fat cells.

hypothalamus   The part of the brain that regulates body temperature, appetite, and hormonal secretions.

ileum   An 11-foot section of the small intestine leading to the large intestine.

implantation   The process by which the blastocyst (the early conceptus produced by rapid cell division of the fertilized egg) burrows into the lining of the uterus. Cells from both the blastocyst and uterus help form the placenta that provides the developing embryo and fetus with nutrients and blood.

inferior   Closer to the soles of the feet. Also known as *caudal*.

inferior vena cava   One of the great vessels of the heart. Returns oxygen-poor blood from the body to the right atrium of the heart.

integumentary system   (after the Greek word meaning "covering") consists of the skin, the hair, and the nails.

intercostal space   The space between each pair of ribs.

intervertebral discs   Located between vertebrae and made of fibrous cartilage.

intervertebral foramen   The hole through the vertebrae through which the spinal cord runs.

intramembranous ossification   The bone-development process in which bone develops before birth within an embryonic connective tissue membrane.

intrinsic hand muscles   Small muscles that originate in the hand. The intrinsic muscles are best suited for delicate, precise movements of the thumb and fingers.

intrinsic muscles of the tongue   Muscles that begin and end in the tongue and do not attach to bone. They change the shape of the tongue.

inversion   Movement of the sole inward, like when someone tries standing on the outsides of their feet.

involuntary muscles   Muscles that are controlled by the autonomic nervous system (such as heart muscle).

jejunum   An eight-foot segment of the small intestine where carbohydrates and proteins are digested.

joint capsule   A sleeve of dense connective tissue fibers in a synovial joint.

joint reduction   The procedure of putting the bones back to their normal anatomic relationship after a dislocation.

keratin   A tough protein that makes skin strong. Continued production of keratin within a skin cell eventually kills it by causing loss of its nucleus and organelles (a process called keratinization).

kidney   One of a pair of bean-shaped organs located against the back wall of the upper abdominal cavity that make urine.

kyphosis   An abnormal or exaggerated thoracic curvature that in severe cases produces a "hump-back" deformity.

laminae   See *vertebral arch.*

large intestine   Approximately five feet long. It consists of the cecum, appendix, ascending colon, transverse colon, descending colon, sigmoid colon, rectum, and the anal canal.

larynx   Commonly called the voice box, it is located in the midline of the neck in front of the pharynx and is connected to the trachea.

lateral   Farther from the midline of the body.

lateral rotation   To move the anterior surface of a limb away from the body. Also known as external rotation.

latissimus dorsi   A shoulder muscle that extends and medially rotates the arm. It is used when a person swims a crawl stroke or chins on a bar.

lens   The part of the eye located behind the iris. It is a transparent, biconvex elastic structure. It changes shape to focus the light rays onto the retina.

leukocytes   White blood cells.

ligament   Fibrous tissue that connects one bone to another.

liver   An abdominal organ that processes and stores nutrients absorbed in the digestive tract, neutralizes and eliminates toxic substances, produces bile and important blood plasma proteins, and serves as a reservoir for certain minerals and fat-soluble vitamins.

loose connective tissue   Found throughout the body. It is best seen in the hypodermis or subcutaneous tissue layer between the skin and muscles. This tissue contains variable amounts of fat cells within a delicate meshwork of fibers. It provides support to epithelial tissue and encloses blood vessels and lymphatic channels.

lordosis   An abnormal or exaggerated lumbar curve that produces a "swayback" deformity.

lower limb   The hip, thigh, leg, and foot (on both sides). It supports the weight of the body and serves as transportation.

lymph   A translucent, straw-colored fluid that is similar to blood plasma, but with a lower concentration of proteins.

lymph node   Small filter consisting of a fine sponge-like meshwork of connective tissue fibers and containing immune cells.

lymphatic capillaries   Thin-walled, closed-ended sacs that drain tissue fluid into larger lymphatic vessels.

lymphocytes   Cells that destroy foreign cells. T-lymphocytes (or T-cells) destroy foreign or abnormal cells, and B-lymphocytes (or B-cells) produce antibodies during an immune response.

matrix   The nonliving substance in which living cells are embedded.

medial   Closer to the midline of the body.

medial rotation   To move the anterior surface of a limb toward the body. Also known as internal rotation.

median or mid-sagittal plane   Divides the body into equal left and right halves. Planes that are parallel to the median plane (also called the midline) are called *paramedian*, *sagittal*, and *parasagittal* planes.

melanin   The most common pigment found in the skin. It serves to darken the skin following sun exposure, to protect it from ultraviolet radiation.

meninges   The coverings of the spinal cord that protect it from injury.

metacarpals   The bones of the palm of the hand.

metatarsals   The five smaller forefoot bones, just before the phalanges.

microvilli   Slender, finger-like projections on the surfaces of cells. They increase the cells" surface area to enhance absorption of nutrients during digestion or transport of fluids and ions during urine formation.

motor areas   Parts of the brain that initiate messages for movement.

mucosa   The lining of hollow organs, such as the small intestine.

muscle fiber   A single cell of a muscle.

muscles of facial expression   Small, delicate muscles that attach from bones of the skull into the overlying skin. They move the skin to create the many expressions of emotion that we recognize as human.

myelin   An insulating material that helps to increase the speed of transmission of nerve impulses along axons.

myocardium   The middle layer of the heart.

nephro-   A Greek word that means "kidney." Both terms are used a lot when describing parts of the kidney.

nephron   The functional unit of the kidney. There are over one million nephrons in each kidney.

nerve tracts   Pathways for sensations to travel up and down the spinal cord.

neurocranium   The floor, sides, and roof of the cranium. The neurocranium contains and protects the brain.

neuroglial (or glial) cells   Cells that support the neurons.

neurons   Also known as nerve cells, which form the structural and functional unit of the nervous system.

nonsynovial joints   Joints with little space between the bones, which allows limited or no movement.

nuclei   Collections of nerve cell bodies located *inside* the central nervous system.

occipitofrontalis   The scalp muscle, which attaches to the skin over the eyebrows. Contraction of this muscle causes the forehead to wrinkle, as in the expression of surprise.

oocytes, or **ova**   The egg cells produced by the female ovaries.

orbicularis oculi   A large circular-shape muscle that surrounds the orbit (the eye socket) and the eyelids.

orbicularis oris   A circular-shape muscle surrounding the mouth.

orbit   The bony socket that contains the eyeball.

os coxae   Hip bones.

ossification   The process of replacing soft cartilage with bone matrix that mineralizes, thus forming real bone. See *endochondral ossification* and *intramembranous ossification.*

osteoarthritis   The most common degenerative disease of the joints.

osteoblasts   Cells that make new bone.

osteoclasts   Cells that break down and reabsorb old bone.

osteocytes   Cells that are part of the bone.

osteocytic lacunae   The cavities that contain osteocytes.

osteoid   An organic material, about 90 percent collagen, that lacks bone minerals.

osteoporosis   A loss of bone mass so severe that people (particularly elderly women) are more likely to break bones during normal activity.

osteoprogenitors   Cells that produce osteoblasts, which in turn form the bone.

ovaries   The female reproductive organs that produce the egg cells. The two ovaries are located in the pelvic cavity next to the uterus and beneath the Fallopian or uterine tubes.

palate   The roof of the mouth.

pancreas   An elongated gland located along the back wall of the upper portion of the abdominal cavity. Its exocrine secretions digest and metabolize food and its endocrine secretions regulate the blood sugar.

paranasal sinuses   Membrane-lined, air-filled spaces within hollowed-out portions of bones bordering the nasal cavity.

parathyroid glands   Small endocrine glands located on the posterior surfaces of the lateral lobes of the thyroid gland, that produce parathyroid hormone. This hormone acts on bone cells for calcium and phosphorus metabolism.

patella   Kneecap bone.

pectoral region   An area of the body that includes the pectoralis major muscle, an upper limb muscle, and overlying it, the breast. This area extends from the clavicle to about the sixth rib and from the breastbone to the armpit.

pectoralis major   A powerful adductor muscle and medial rotator of the arm.

pedicles   See *vertebral arch.*

pelvic diaphragm   A funnel-shaped muscular floor. This thin diaphragm consists of skeletal muscle that attaches to the inner walls of the pelvic cavity from the pubic bone to the coccyx.

pelvic inlet   Also known as the pelvic brim, this is the upper opening of the pelvis.

pelvic outlet   A diamond-shaped area formed by bone and ligaments.

pelvis   The region made up of the two hip bones and the sacrum; it protects the pelvic organs and transfers the weight of the body to the lower limbs.

penis   The main organ of the male reproductive system, it carries urine from the bladder out of the body. It is also used in sexual intercourse to transfer sperm to the vagina.

perineum   The area of the body below the pelvic diaphragm and between the thighs. In this region are the external genitalia, the urethra, and the anal canal.

periosteum   A connective tissue covering on the surface of the bone that contains osteoblasts, nerves, and blood vessels.

peripheral nervous system Consists of the 31 pairs of spinal nerves, so called because they arise from the spinal cord, and the 12 pairs of cranial nerves that originate from the brain. The PNS relays information to and from the body to the brain and spinal cord and is functionally divided into the somatic nervous system and the autonomic nervous system.

peristalsis The rhythmic contraction of smooth muscle that propels food through the digestive tract and lymph toward the lymph nodes.

phalanges The bones of the fingers. Also the bones of the toes.

pharynx A funnel-shape muscular tube that attaches to the base of the skull and connects the nasal cavity to the esophagus in the lower neck.

physiological sphincter A valve that lacks a distinctive ring of smooth muscle but is closed until autonomic nerves cause it to relax and open when needed.

pia mater The innermost membrane covering of the meninges.

plantar flexion Movement of the toes and foot toward the ground, like when someone stands on their toes.

pleura The membranes that cover the lungs and line the pulmonary cavity. The word is derived from a Greek word meaning "rib or side."

plexus In the nervous system, a network of ventral primary rami from several spinal cord levels that join to form definitive nerves.

popliteal fossa The diamond-shape region behind the knee bordered by several muscles.

portal vein A large vein that is formed when the superior mesenteric vein joins the splenic vein.

posterior A structure that posterior (or dorsal), is closer to the back of the body than a structure that is anterior (or ventral).

posterior compartment Located on the backside of the bones in the arm, forearm, thigh, and leg.

primary curves Thoracic and sacral spinal curves that face the front of the body and are formed before birth.

primary trabeculae Narrow beams of mineralized cartilage left behind by the invasion of blood vessels and osteoclasts.

pronation Medial rotation of the forearm so that the palms face backward (posteriorly).

prostate A walnut-shaped gland surrounding the male urethra. It produces some of the seminal fluid.

proximal A term used often in anatomy to refer to a structure that's closest to the center of the body or other point of reference.

pseudostratified epithelium Tissue that lines the passageways of parts of the nasal cavity, trachea, and bronchi. Also called *respiratory epithelium*. Although this epithelium looks like it's made up of several layers of cells, it actually consists of a single layer of cells with different shapes and sizes.

pulmonary trunk One of the great vessels of the heart. It divides into the right and left pulmonary arteries that carry venous blood from the right ventricle to the lungs, where it becomes oxygenated.

pulmonary veins Carry oxygenated blood from the lungs to the left atrium.

pyloric sphincter A valve that controls passage of gastric contents into the small intestine.

radius One of the two bones of the forearm.

rectum A storage organ that accumulates and holds the fecal mass until it passes out the anal canal and anus during defecation.

renal A Latin word that refers to anything of or near the kidneys.

respiration The process of breathing, in which carbon dioxide is taken from blood cells and replaced with oxygen.

respiratory diaphragm The mobile fibro-muscular partition separating the thoracic and abdominopelvic cavities. Its contraction increases the volume of the thoracic cavity and allows air to enter the lungs during breathing.

reticular fibers Consist of collagen and are much thinner than regular collagen fibers. They serve as a delicate network around nerves and smooth muscle. They also form a mesh-like framework for hemopoietic and lymphoid organs, such as bone marrow, spleen, and lymph nodes.

retroperitoneal Refers to organs that lie behind the parietal peritoneum, including the kidneys and ureters, and the adrenal glands.

rotator cuff The group of muscles that stabilize and move the arm. Includes the supraspinatus, infraspinatus, teres minor, and subscapularis.

sacral canal A hole within the sacrum that aligns with the vertebral canal and contains the lower end of the dural sac with its sacral and coccygeal nerve roots.

sacral hiatus Is the opening at the end of the sacral canal.

sacroiliac The joint where the sacrum joins the hip bone.

sacrum A single triangular bone in the pelvis consisting of five vertebrae fused together.

salivary glands Three pairs of glands in the facial area that produce saliva and aid in diges-tion of food.

scapula Shoulder blade; part of the appen-dicular skeleton.

scoliosis An abnormal side curvature of the spine.

scrotum The pouch-like sac of skin that holds the testicles outside the body.

sebaceous glands Glands located throughout the body, but not in the skin of the palms of the hands or soles of the feet. Sebaceous glands are simple sac-like glands with a duct that opens to the upper end of a hair follicle. Cells in these glands produce an oily substance.

secondary curves Cervical and lumbar spinal curves that face the rear of the body and form as people grow and develop.

semen Fluids produced by the accessory geni-tal glands that activate sperm cells and add vol-ume to the ejaculate.

seminal vesicles Two narrow, sac-like struc-tures about 5-6 inches in length located behind the bladder. They produce semen.

sensory areas Parts of the brain that receive signals from the spinal cord and brain stem regarding touch, taste, smell, sight, and hearing.

serratus anterior An important muscle in protracting the scapula, as when a person wants to increase his or her forward reach, and in rotating the scapula upward.

simple columnar epithelia Epithelial tissue in which cells are taller than they are wide. This kind of lining is most prevalent in many of the organs of the digestive system and in the female reproductive system.

simple cuboidal epithelia Simple layers of cells that look box-like. They commonly line portions of ducts of glands and tubules in the kidney.

simple squamous epithelia Epithelial tissue that consists of a single row of very thin or flat-tened cells.

skeletal muscles The voluntary muscles attached to bones that help the body move.

smooth muscle A muscle type found in the walls of the hollow organs (for example, the digestive, urinary, and reproductive systems) and the blood vessels. Smooth muscle is also found in the iris of the eyeball, where it controls the diameter of the pupil, as well as in the skin, where it attaches to hair follicles to produce goose bumps in response to cold temperatures or fear.

somatic nervous system   A functional division of the peripheral nervous system that regulates the body's external environment and causes skeletal muscle to contract.

spermatogenesis   The production of sperm cells.

spinal nerve   One of the 31 pairs of nerves attached to the spinal cord.

spinous process   A single bony part of a vertebra located where two laminae meet.

spleen   A soft, fist-size organ located in the left upper quadrant of the abdomen to the left of the stomach and next to the rib cage. It is the largest of the body's lymphoid organs and responds to antigens in the blood stream.

sternum   Breastbone.

stratified epithelia   Epithelial tissue that has several rows or layers of cells.

subarachnoid space   A space in the meninges that contains cerebrospinal fluid, which cushions and protects the brain and the spinal cord within the dural sac.

sulci   Grooves on the surface of the heart and the brain.

superficial   Closer to the surface or skin of the body.

superior   Closer to the head or cranium. Also known as *cranial*.

superior vena cava   One of the great vessels of the heart. Returns oxygen-poor blood from the upper body and thorax to the right atrium of the heart.

supination   Lateral rotation of the forearm so that the palms face forward (anteriorly).

sutural joints   Joints between the bones of the skull.

synapses   The spaces between a nerve cell and another cell, through which impulses are transmitted.

synarthroses   Another term for nonsynovial joints.

syndesmoses   A type of fibrous joint consisting of ligaments that connect two bones, such as the interosseous membranes that connect the bones of the forearm and the leg.

synovial fluid   An egg white-like fluid that covers the cartilage in the joint to facilitate smooth movement and keeps it healthy.

synovial joints   Joints with a space between the bones, which contains synovial fluid.

synovial membrane   Soft tissue that lines the joint capsule.

systole   The period of ventricular contraction in the heartbeat.

talocrural joint   The ankle joint.

tarsals   The seven larger hind-foot bones nearest the ankle.

testes (or testicles)   Soft, oval structures inside a pouch-like sac of skin called the scrotum. Sperm production occurs inside the testes.

testosterone   The principal male sex hormone.

thalamus   The part of the brain that acts as a translator between various parts of the brain. It also regulates states of sleep and wakefulness.

thoracic duct   The largest lymphatic vessel in the body.

thoracic wall   Consists of 12 thoracic vertebrae in back, the sternum in front, and 12 pairs of ribs in between.

thoracolumbar outflow   A term that describes the sympathetic division of the nervous system, by reference to the location of the cell bodies of its preganglionic neurons in lateral horns of the T1 to L2 segments of the spinal cord.

thorax   The part of the body between the neck and the abdomen.

thymus   A primary lymphoid organ with two lobes, located in the upper chest behind the sternum.

thyroid gland   An endocrine gland that secretes thyroxine for regulating metabolic activities in the body and calcitonin for regulation of blood calcium levels.

tibia   The larger of the two bones of the leg.

tissue   Similar cells working together to support a particular purpose or function.

trabecular or **cancellous bone**   Spongy tissue at the ends of long bones.

trachea (or windpipe)   Connected to the lower end of the larynx. It passes from the neck into the upper part of the thorax before dividing into right and left bronchi.

transitional epithelium   The epithelial lining of the urinary passages and bladder. This tissue must be able to expand to temporarily hold the urine produced by the kidneys. The outermost cells are shaped like cubes when the bladder is empty. They flatten as the bladder fills.

transverse foramina   The holes in the transverse processes of the neck vertebrae.

transverse processes   A pair of bony parts on vertebrae at the junction between a pedicle and a lamina.

trapezius   The largest muscle of the shoulder, which moves the scapula.

triceps brachii   A muscle in the posterior compartment of the arm that has three heads: the long head, the lateral head, and the medial head. Together these connect the scapula and humerus to the ulna and extend the forearm.

true pelvis   The area between the pelvic inlet and the pelvic outlet.

tubule   A thin, elongated channel or tube in the body used to conduct cells or fluids.

ulna   One of the two bones of the forearm.

upper limb   The shoulder, arm, forearm, and hand.

ureter   One of the two tubes that connect the kidneys to the bladder.

urethra   The passageway that conducts urine from the bladder out of the body. In males, semen also passes through the urethra.

uterine (fallopian) tubes   Narrow, muscular tube structures that extend from the uterus to each ovary. Each tube is about five inches long and conducts the egg from the ovary to the uterus.

uterus, or **womb**   A hollow, muscular organ that looks like an upside-down pear. It is about three inches long, two inches wide, and an inch thick. The uterus is located in the center of the pelvic cavity behind the urinary bladder and in front of the rectum. It is where the developing fetus lives and grows.

vagina   A hollow, distensible fibromuscular structure about three to five inches in length. It opens at one end with the vestibule, and surrounds and attaches at its opposite end to the cervix of the uterus.

valves   Located at the openings between the four chambers of the heart. There are also valves between the chambers and the arteries leaving the heart. They control the direction the blood flows within and out of the heart.

venous blood   Blood from the body's veins that returns to the heart before being pumped to the lungs for oxygenation.

ventral   See *anterior.*

ventral horns   Are the parts of the gray matter of the spinal cord that contain somatic motor nerves whose axons send impulses to skeletal muscles.

ventral rami   Branches of spinal nerves that provide sensory and motor supply to skin and muscles of the body wall and upper and lower limbs.

ventricles (brain)   Interconnected, fluid-filled chambers in the brain that produce and hold cerebrospinal fluid.

ventricles (heart)   The lower chambers of the heart, left and right.

vertebra prominens   A long spinous process on the C7 vertebra. It's the first bump you can feel at the bottom of your neck.

vertebrae   Irregular-shaped bones that make up the spine (and when you're talking about just one of them, it's a *vertebra*). Each spine has 33 of them. They are divided into five regions: cervical (neck), thoracic (connected to the ribs), lumbar (lower back), sacral (part of the pelvis), and coccygeal (tailbone).

vertebral arch   Made up of pedicles and laminae, forming a hole called the vertebral foramen, which houses the spinal cord.

vertebral bodies   The main weight-bearing parts of the vertebrae.

vertebral column   A semi-rigid structure consisting of a block-like series of bones, called vertebrae, connected by discs, ligaments, joints, and muscles. Also known as the spinal column, spine, or backbone.

vertebral notches   Located on the respective upper (superior) and lower (inferior) surfaces of the pedicles. By placing two vertebrae on top of one other, these notches form a hole called the intervertebral foramen.

villi   Microscopic finger-like extensions that project into the passageway of the small intestine.

viscerocranium   Also known as the facial skeleton. It contains and protects the upper parts of the digestive and respiratory systems.

vitreous body   Located behind the lens. It consists of a transparent jelly-like fluid, the vitreous humor, that fills the vitreous chamber between the lens and retina and gives the eyeball its shape.

voluntary muscles   Muscles that are consciously controlled by the person (for example, the skeletal muscles that move the arm).

Further Reading

Books

Alcamo, I. Edward, and Barbara Krumhardt. *Anatomy and Physiology the Easy Way*. Hauppauge, NY: Barron's Educational Series, 2005.

Anatomical Chart Company. *Rapid Review: Anatomy Reference Guide*. Skokie, IL: Lippincott Williams and Wilkins, 2005.

Biel, Andrew, and Robin Dorn (illustrator). *Trail Guide to the Body: How to Locate Muscles, Bones, and More*. Boulder, CO: Books of Discovery, 2005.

DK Publishing. *The Human Body: An Illustrated Guide to Its Structure, Function, and Disorders*. New York: DK Books, 1995.

Drake, Richard, Wayne Vogl, and Adam M. W. Mitchell. *Gray's Anatomy for Students*. Philadelphia: Churchill Livingstone, 2004.

Dudek, Ronald W., and Thomas M. Louis. *High-Yield Gross Anatomy*. Philadelphia: Lippincott Williams and Wilkins, 2007.

Goldberg, Stephen, and Hugue Ouellette. *Clinical Anatomy Made Ridiculously Simple*. Miami: MedMaster, 2007.

Gray, Henry. *Gray's Anatomy*, Classic Collector's Edition. New York: Gramercy, 1988.

Kapit, Wynn, and Lawrence Elson. *The Anatomy Coloring Book*. San Francisco: Benjamin Cummings, 2001.

Moore, Keith L., and Arthur F. Dalley II. *Clinically Oriented Anatomy*, Fifth Edition. Philadelphia: Lippincott Williams and Wilkins, 2005.

Parker, Steve. *The Human Body Book*. New York: DK Books, 2007.

Rosse, Cornelius, and Penelope Gaddum-Rosse. *Hollinshead's Textbook of Anatomy*. Philadelphia: Lippincott-Raven Publishers, 1997.

Shmaefsky, Brian R. *Applied Anatomy and Physiology*. St. Paul, MN: Paradigm Publishing, 2007.

Van De Graaf, Kent, and R. Ward Rhees. *Schaum's Outline of Human Anatomy and Physiology*. New York: McGraw-Hill, 2001.

Websites

academic.pgcc.edu/~aimholtz/AandP/ AandPLinks/ANPlinks.html
A massive collection of helpful links from Prince George Community College.

www.anatomy.org/
Official site of the American Association of Anatomists.

anatquest.nlm.nih.gov/
AnatQuest is a database of anatomical images.

www.bartleby.com/107/indexillus.html
Bartleby.com offers the original 1918 edition of *Gray's Anatomy* online.

www.bio.psu.edu/people/faculty/strauss/ anatomy/biology29.htm
Dr. James Strauss of Penn State provides extensive anatomical photos.

www.bodyworlds.com/en.html
Information on exhibits of "plastinized" human bodies.

www.dartmouth.edu/~anatomy/
Human anatomy learning modules from Dartmouth.

ect.downstate.edu/courseware/haonline/ index.htm
An interactive dissection tutorial.

www.innerbody.com/htm/body.html
A tour of the body's systems.

www.instantanatomy.net
A website featuring beautifully colored anatomic drawings with labels, podcasts, and PowerPoint audiovisual lectures.

www.ithaca.edu/faculty/lahr/LE2000/LE_ index.html
Interactive dissection site with cadaver photos, from Ithaca College.

www.med.umich.edu/lrc/coursepages/M1/ anatomy/html
A rich resource of anatomic and radiographic images, tables, quizzes, anatomy crossword puzzles, and other learning games.

www.nlm.nih.gov/medlineplus/anatomy.html
MedLine Plus is a service of the U.S. National Library of Medicine and the National Institutes of Health. This page has an amazing collection of links to overviews, pictures, tutorials, research, and more.

www.nlm.nih.gov/research/visible/ visible_human.html
The Visible Human Project created complete, anatomically detailed 3-D representations of male and female human bodies using cadavers that were microscopically sliced and photographed.

www.sickkids.ca/childphysiology/cpwp/ ChildPhysiologyhome.htm
A website for parents to learn about their children's bodies.

Magazines and Journals

American Heart Journal (journals.elsevierhealth.com/periodicals/ymhj)

American Journal of Nursing (ajnonline.com)

American Journal of Physiology—Endocrinology and Metabolism (ajpendo.physiology.org/)

American Journal of Surgery (www.sciencedirect.com/science/journal/00029610)

The Anatomical Record (www.anatomy.org)

Blood (bloodjournal.hematologylibrary.org)

Bone: The Official Journal of the International Bone and Mineral Society (www.elsevier.com/wps/find/journaldescription.cws_home/525233/description)

Bone and Joint Letter (www.ovid.com/site/catalog/Journal/1873.jsp)

BRAIN: A Journal of Neurology (brain.oxfordjournals.org)

Cell (www.cell.com)

Clinical Anatomy (www.clinicalanatomy.org/journal.html)

Gastroenterology: Official Journal of the AGA Institute (www.gastrojournal.org/)

Journal of Anatomy (www.blackwellpublishing.com/journal.asp?ref=0021-8782&site=1)

Journal of Chemical Neuroanatomy (www.elsevier.com/wps/find/journaldescription.cws_home/524994/description#description)

Journal of Musculoskeletal Medicine (www.amazon.com/Journal-of-Musculoskeletal-Medicine/dp/B00006KKIJ)

Journal of the American Academy of Physician Assistants (www.jaapa.com/)

Journal of the American Medical Association (jama.ama-assn.org)

Journal of Urology (www.jurology.com)

The Lancet (www.thelancet.com/journals)

Neuroanatomy (www.neuroanatomy.org)

Obstetrics and Gynecology (www.greenjournal.org)

Spine (www.spinejournal.com)

Surgical and Radiologic Anatomy (www.springerlink.com/content/100112/)

Thorax: An International Journal of Respiratory Medicine (thorax.bmj.com)

Index

A

abdomen, 215-225
abducens nerve (CN VI), 264
abduction
   adduction, compared, 154
   fingers, 194
abnormal curves, spine, 160-162
accessory organs (digestive
  system), 118-121, 221-223
acetabulum, 228
acid reflux, 112
acromioclavicular joint, 181
acromion process, 178
adduction
   abduction, compared, 154
   fingers, 194
adipocytes, connective tissue, 15
adrenal glands, 123-225
adventitia, heart, 84
afferent arterioles, 126
albinism, 30
alveolar ducts, 105
alveoli, 105
amylase, 110
anal canal, 236-237
anal triangle, 236-237
anatomic sphincter, stomach,
  113
anatomical movement, 154-157
anatomical planes, 152-153
anatomical position, 151-153

anatomical relationships,
  152-153
anatomical terms of relationship
  or position, 153
ankle, 247-249
annular ligament, 183
ANS (autonomic nervous system), 61-63, 69-71
   parasympathetic system,
    69-71
   spinal nerves, 65-67
   sympathetic system, 69-70
anterior abdominal wall,
  215-218
anterior compartment, leg, 188,
  246
anterior compartment, thigh,
  243-244
anterior cruciate ligaments, 250
anterior lobe, prostate gland,
  232
anterior longitudinal ligament,
  170
anterior nasal aperture, 266
anterior parts, 152
anterior view, pelvis, 228
antibody-mediated immunity, 93
antrum, 113
anulus fibrosus, 169
aortic semilunar valves, 81
apocrine sweat glands, 33
aponeuroses, 217

appendicular system (skeletal),
  39-41
appendix, 94
aqueous humor, eyeball, 265
arachnoid mater, meniges, 55
arcuate line of the ilium, 229
arm
   bones, 182-183
   muscles, 188-189
arteries
   elastic arteries, 84
   lower limb, 252
   upper limb, 196
articular cartilage, 43
articular processes, vertebrae,
  164
articulations, 41-43
ascending pathways, central
  nervous system, 49-50
association areas, lobes, 48
asthma, 104
astrocytes, 22
atlas, vertebra, 165
atrioventricular valves (AVs), 81
auditory ossicles, 273
auditory tubes, 273
autonomic nervous system
  (ANS), 61-63
AVs (atrioventricular valves), 81
axial system (skeletal), 39-40
axilla, 194-199
axillary artery, 194-195

axillary nerve, 197
axis, vertebra, 165
axons, neurons, 20
azygous system of veins, 206

B

B-lymphocytes, 91
back, 159
    deep back muscles, 172-173
    meninges, 173-174
    spinal cord, 173-174
    spinal nerves, 173-174
    spine, 159
        curves, 160-162
        regions, 160
        vertebrae, 159-169
    superficial back muscles, 172
    vertebral column, 159
        ligaments, 170-171
    vertebral joint, 169-170
back muscles
    deep back muscles, 172-173
    superficial back muscles, 172
back parts, 152
basale stratum, epidermis, 29
basal ganglia, role, 51
basal nuclei, 47
Bell's palsy, 262
belly, 215
    accessory organs, 221-223
    anterior abdominal wall,
      215-218
    GI (gastrointestinal) tract,
      218-221
        esophagus, 218-219
        large intestine, 220-221
        small intestine, 220
        spleen, 220
        stomach, 220
    posterior abdominal wall,
      224-225
benign prostatic hypertrophy
  (BPH), 232

bicipital groove, 180
bicuspid valve, 81
bile, 121
bile canaliculi, liver, 120
blockages, ureters, 230
blood-brain barriers, 22
blood supply
    lower limb, 251-253
    upper limb, 194-197
blood vessels, 84-85
bodies, vertebrae, 163
bones, 35-36
    cancellous bones, 39
    cells, 36
    cortical bones, 38
    flat bones, 38
    growth and development,
      36-39
    irregular bones, 38
    joints, 41-43
    long bones, 38
    lower limb, 239-242
        ankle, 247-249
        foot, 247-249
        gluteal region, 240-242
        hip, 240-242
        leg, 245-247
        thigh, 242-245
    pelvic region, 227-228
    short bones, 38
    skull, openings, 257-259
    upper limb, 178
        arms, 182-183
        elbow, 183-184
        forearms, 182-183
        hand, 184-186
        shoulder region, 178-181
        wrist, 184-185
bony labyrinth, 274
Bowman's capsule, 126
BPH (benign prostatic hypertrophy), 232
brachial plexus, 197
brachiocephalic trunk, 209
brain, 45-46
    cerebrum, 47-51

CSF (cerebrospinal fluid),
      56-59
    sagittal view, 46
    ventricles, 58
brain stem, 52-53
brachial plexus, 69
breastbone, 178
breast region, 202-204
breasts (female), mammary
  glands, 145-147
breathing mechanics, 212-213
broad ligament, 234
bronchi, 103-104
bronchioles, 104
bronchopulmonary segments,
  210
buccinator, 262
bulbourethral glands, 135-136
bulbs, hair, 31

C

canaliculi, 36
cancellous bones, 39
capillaries, 85
    lymphatic capillaries, 89-90
capitulum, 183
cardiac muscle, 18-19
cardiac tamponade, 207
cardiovascular system
    blood vessels, 84-85
    capillaries, 85
    heart, 80-83
    lymphatic system, 89-92
    lymphoid organs, 92-95
    portal venous system, 87
    systemic arteries, 83
    veins, 86
caridac region, stomach, 113
carpal bones, 38
carpal tunnel, muscles, 194-195
carpal tunnel syndrome, 194
carpometacarpal (CMC) joints,
  185
cartilaginous joints, 41-42

CAT (computed tomography)
scans, 8
cauda equina, 173
cell-mediated immunity, 93
cell biology, 8
cell body, neurons, 20
cells
 B-lymphocytes, 91
 bones, 36
 chondroblasts, 37
 connective tissue, 14-15
 glial cells, 22
 lymphocytes, 91
 macrophages, 91
 T-lymphocytes, 91
central lymphoid organs, 92
central nervous system, 45
 ascending pathways, 49-50
 brain, 45-46
  brain stem, 52-53
  cerebellum, 52
  cerebrum, 47-51
  diencephalon, 51-52
  sagittal view, 46
 descending pathways, 49-50
 meniges, 55-57
 spinal cord, 54-55
 ventricular system, 58-59
central process, 64
central sulcus, 48
cerebellar nuclei, 52
cerebellum, 52
cerebral aqueduct, 58
cerebral hemispheres, brain, 47
cerebrospinal fluid (CSF), 55-59
cerebrum, 47-51
cervical curve, spine, 160
cervical pleura, 209
cervical plexus, 69
cervical triangles, 274
cervical vertebrae, 164-165
cervical viscera, 275
cervix, 232
 uterus, 143
chewing, 111
 muscles, 269

chondroblasts, 15, 37
chyme, stomach, 112
circulatory system. *See* cardiovascular system
circumduction, 155
circumvallate, 270
cisterna chyli, 91, 117
clavicle, 178
clitoris, 144
CMC (carpometacarpal) joints,
185
CN I (olfactory nerve), 67
CN IV (trochlear nerve), 67
CN IX (glossopharyngeal
nerve), 67
CN VII (facial nerve), 67
coastal, 205
coccygeus, 229
coccyx, 169
collagen, connective tissue, 15
common bile duct, 121
common fibular nerve, 254
computed tomography (CAT
scans), 8
concave regions, kidneys, 124
conchae, 98
condyles, 183
connective tissue, 14-17
coracoid process, 180
cornea, eyeball, 265
corneal reflex, 264
corneum stratum, epidermis, 28
coronal plane, 152
coronoid process, 183
corpora cavernosa, 137
corpus collosum, cerebrum, 47
corpus spongiosum, 137
cortical bones, 38
costal arch, 204
costal cartilage, 204
costal margins, 204
costal pleura, 209
cranial nerves, 67-68, 260
craniosacral outflow, parasympathetic outflow, 70
cremasteric muscle, 134

cricothyroid membrane, 275
cricothyrotomies, 275
cryptochidism, 132
CSF (cerebrospinal fluid),
55-59, 173
cubital fossa, 196
cupola, 209
curvatures, stomach, 113
curves, spine, 160-162

D

deep back muscles, 172-173
deep brachial artery, 196
deep femoral artery, 251
deep versus superficial, 153
De Humani Corporis Fabrica, 7
deltoid, 187
deltoid tuberosity, 181
dendrites, neurons, 20
dense connective tissue, 16-17
denticulate ligaments, 56
dermal papilla, 31
dermis, 28-31
descending pathways, central
nervous system, 49-50
development, bones, 36-39
developmental anatomy, 8
diaphragmatic pleura, 209
diaphysis, 37
diarthrodial joints, 42
diastole, 81
diencephalon, 51-52
digestive system, 109
 accessory organs, 118-121,
 221-223
 esophagus, 111-112
 large intestine, 118
 mouth, 109-111
 pharynx, 111-112
 small intestine, 114-117
 stomach, 112-113
dislocations, joints, 182
distal radioulnar joint, 184
distal versus proximal, 153

diverticuli, 221
Dorsal interossei ABduct the fingers (DAB), 193
dorsalis pedis artery, 251
dorsal primary ramus, 64
dorsiflexion versus plantar flexion, 156
ductus deferens, 133-134
duodenum, small intestine, 114-116
dural sac, 55-56, 167
dura mater, meniges, 55

E

ears, 272-274
eccrine sweat glands, 33
efferent fibers, 63
ejaculatory ducts, 134, 232
elastic arteries, 84
elastic fibers, connective tissue, 15
elbow bones, 183-184
embryology, 8
endocardium layer, heart, 80
endochondral ossification, 37
endocrine pancreas, 119
endometrium, uterus, 143
ependyma, 58
epicardium layer, heart, 80
epidermis, 28-30
epididymis, 133-134
epiploic appendages, 221
epithelial tissue, 11-14
Erasistratus, 6
erector spinae, 173
esophagus, 111-112, 218-219, 275
ethmoid sinuses, 266
eversion versus inversion, 156
exocrine pancreas, 119
extension versus flexion, 154
extensor compartment, 188-191
extensor retinaculum, 189
external abdominal oblique, 217

external genitalia (female), vagina, 144-145
external rotation, 154-155
extraocular muscles, eye, 263-264
extrinsic muscles
    hand, 192
    tongue, 269
eyeball, 262-265

F

face muscles, 261-262
facial nerve (CN VII), 67
fallopian tubes, 141
false pelvis, 229
false vocal cords, 101
fascia lata, 243
female pelvic organs, 232-234
female perineum, 235
female reproductive system, 139
    external genitalia, 144-145
    internal reproductive organs, 139-145
    mammaries, 145-147
female urethra, 128
femoral artery, 251
femoral nerve, 253
femur, 243
fibers
    connective tissue, 15-16
    efferent fibers, 63
    muscle fibers, 17
fibroblasts, connective tissue, 15
fibrous cartilage, 42, 163
fibrous joints, 41-42
fibrous pericardium, 207
fibular nerve, 254
filiform, 270
filum terminale, 56
fimbriae, 141, 233
fingernails, 32
fingers, 194
fissures, 52
fissures, brain, 47

flat bones, 38
flexion, extension, compared, 154
flexor compartment, 188-190
folia, 52
foliate, 270
follicles, 31, 140
follicle stimulating hormone (FSH), 140
follicular cells, 140
foot, 247-249
    processes, 22
foramen, 163
forearm
    bones, 182-183
    muscles, 189-192
foreskin, penis, 136
fossae, 178
front parts, 152
frontal lobes, cerebrum, 48
frontal plane, 152
frontal sinus, 266
front wall, abdomen, 215-218
FSH (follicle stimulating hormone), 140
fundus, 232
fundus,
    stomach, 113
    uterus, 143
fungiform, 270

G

Galen, 6-7
gallbladder, 121, 223
ganglia, 47
gaseous exchange, respitory system, 107
gastric, 113
gastric glands, 113
gastric pits, 113
genioglossus, 270
GERD (gastroesophageal reflux disease), 112
germinative layer, epidermis, 29

GI (gastrointestinal) tract, 218-221
glands
    adrenal glands, 123, 225
    male reproductive system, 135-136
    mammary glands, 145-147
    parathyroid glands, 275
    prostate gland, lobes, 232
    salivary glands, 110-111, 271
    sebaceous glands, 32-33
    stomach, 113
    suprarenal glands, 123
    sweat glands, 32-33
    thyroid gland, 275
glenohumeral joint, 181
glenoid cavity, 180
glial cells, nervous tissue, 22
glomerular filtrate, 127
glomerulus, 126
glossopharyngeal nerve (CN IX), 67
gluteal region, 240-242
gluteus maximus, 241
gluteus medius, 241
goblet cells, 13
granular cell layer, epidermis, 29
granulosum stratum, epidermis, 29
gray matter
    brain stem, 52
    cerebrum, 47
greater curvature, stomach, 113
greater trochanters, 242
greater tubercles, 180
gross anatomy, 8
ground substance, connective tissue, 16
growth, bones, 36-39
gyri, brain, 47

H

hair bulbs, 31
hair follicles, 31

hairs, 31-32
hamstrings, pulled hamstrings, 245
hands
    bones, 184-186
    joints, 184-186
    muscles, 192-194
haustra, 221
head
    ears, 272-274
    eyeball, 262-265
    face muscles, 261-262
    mouth, 268-271
    nasal cavity, 266-268
    nose, 266-268
    orbit, 262-265
    salivary glands, 268-271
    scalp muscles, 261-262
    skull openings, 257-259
    tongue, 268-271
hearing (sense of), 272-274
heart, 207-209
    adventitia, 84
    intima, 84
    layers, 80
    left margin of the heart, 208
    lower margin of the heart, 208
    media, 84
    pericardium, 207
    right margin of the heart, 208
    superior border of the heart, 209
    valves, 80-82
    vessels, 82-83
hepatic, 120
hepatic ducts, liver, 120
hepatopancreatic ampulla, 220
Herophilus, 6
hilum, kidneys, 124
hip, 240-242, 249
hip bone, 239
histology, 7-8
horizontal plane, 152
hormones, 133
humeroradial articulation, 183

humeroulnar articulation, 183
humerus, 178-180
humoral-mediated immunity, 93
Huntington's disease, 51
hyaluronic acid, 43
hydrocephalus, 46, 59
hyoglossus, 270
hypodermis, 28
hypothalamus, diencephalon, 51-52

I

ileocecal valve, 117
ileum, small intestine, 117
iliotibial band, 243
ilium, 228-229
immune system, 92-95
implantation, 143
incus, ear, 273
inferior thoracic aperture, 205
inferior vena cava, 86
inferior wall, orbit, 263
infraspinatus, 188
infundibulum, 141, 233
inguinal canal, 132
inner medulla region, kidneys, 124
inner oblique layer, stomach, 113
inner wall, orbit, 263
integumentary system, 27
    hairs, 31-32
    nails, 32
    sebaceous glands, 32-33
    skin, 28-31
    sweat glands, 32-33
intercalated discs, 18
intercoastal spaces, 205-206
intercondylar eminence, 245
intercondylar fossa, 242
intercostal nerves, 206
internal abdominal oblique, 217
internal reproductive organs (female), 139-144

internal rotation, 154-155
internal urethral sphincter, 230
interosseous artery, 196
interosseous membrane, 183
interphalangeal (IP) joints, 186
interspinous ligament, 170
intertubercular, 180
interventricular foramina, 58
intervertebral discs, 163,
  169-170
intervertebral foramen, 164
intima, heart, 84
intramembranous ossification,
  36-37
intrinsic muscles
  hand, 192-193
  tongue, 269
inversion, eversion, compared,
  156
involuntary muscles, 19
IP (interphalangeal) joints, 186
irregular bones, 38
ischial tuberosities, 229
ischium, 228
islets of Langerhans, 223

J

jaw joint (TMJ), 257
jejunum, small intestine, 117
joint reduction, 182
joints, 41
  dislocations, 182
  lower limb, 249-251
  nonsynovial joints, 41-42
  sacroiliac joint, 167
  synovial joints, 42-43
  upper limb, 178
    hand, 184-186
    radioulnar joints, 183-184
    shoulder region, 181-182
    wrist, 184-185
  vertebral joints, 169-170

K

keratin, 29
keratinocytes, 29
kidneys, 123-127
kidney stones, 127
knee joints, 250-251
kyphosis, 160-162

L

labia majora, 144
labia minora, 144
laminae, 163
large intestine, 118, 221
laryngopharynx, 100, 275
larynx, 100-102, 275
lateral circumflex femoral arteries, 251
lateral collateral ligament, 247
lateral compartment, leg, 247
lateral condyles, 242
lateral epicondyle, 183
lateral lobes, prostate gland, 232
lateral masses, 165
lateral rotation, 154-155
lateral sulcus, 48
lateral ventricles, brain, 58
lateral versus medial, 153
lateral wall, nasal cavity, 99
lateral wall, orbit, 263
latissimus dorsi, 187
layers
  epidermis, 28
  eyeball, 264
  heart, 80
  stomach, 113
left common carotid artery, 209
left margin of the heart, 208
left subclavian artery, 209
leg, 245-247
lens, eyeball, 265
lesser curvature, stomach, 113
lesser trochanters, 242

lesser tubercles, 180
leukocytes, connective tissue, 15
levator ani, 229
ligaments, vertebral column,
  170-171
ligamentum flavum, 170
liver, 119, 222
lobar bronchi, 103
lobes
  association areas, 48
  cerebrum, 48
  motor areas, 48
  prostate gland, 232
  sensory areas, 48
locations, structures, 153
long bones, 38
longitudinal cerebral fissures,
  cerebrum, 47
loose connective tissue, 16
lordosis, 161-162
lower esophageal sphincter,
  stomach, 113
lower inner (breast), 203
lower jaw bone, 257
lower limb, 242
  blood supply, 251-253
  bones, 239-242
  joints, 249-251
  nerve supply, 253-254
lower margin of the heart, 208
lower outer (breast), 203
lower respiratory tree, 102
  bronchi, 103-104
  lungs, 104-106
  trachea, 103-104
lucidum stratum, epidermis, 28
lumbar curve, spine, 160
lumbar vertebrae, 166-167
lumbosacral plexus, 69
lumen, 111
lungs, 104-106, 209-211
  breathing mechanics, 212-213
lunula, 32
lymphatic capillaries, 89-90
lymphatic channels, 134

lymphatic drainage, breast region, 203-204
lymphatic system, 89-91
capillaries, 89-90
lymph nodes, 90
lymphoid organs, 92-95
lymphs, 90
lymph trunks, 91
thoracic duct, 92
lymphedema, 90
lymph nodes, 90-93
lymphocytes, 91
lymphoid cells, 94
lymphoid organs, 92-95
lymphs, 90
lymph trunks, 91
lysozyme, 110

M

macrophages, 91
connective tissue, 15
magnetic resonance imaging (MRI), 8
male pelvic organs, 231-232
male reproductive system, 131
acceossory glands, 135-136
ductus deferens, 133-134
epididymis, 133-134
penis, 136-137
scrotum, 131-132
testicles, 131-133
male urethra, 129
malleus, ear, 273
mammary glands, 145-147
mandible, 257
mast cells, connective tissue, 15
mastication, 269
maxillary sinus, 266
MCP (metacarpophalangeal) joints, 186
mechanics, breathing, 212-213
media, heart, 84
medial circumflex femoral arteries, 251

medial compartment, thigh, 243
medial compartments, 244
medial condyles, 242
medial epicondyle, 183
medial rotation, 154-155
medial versus lateral, 153
medial wall, orbit, 263
medial wall of the nasal cavity, 98
median lobe, prostate gland, 232
median nerve, 198
median plane, 152
mediastinal pleura, 209
medulla oblongata, brain stem, 53
melanin, 30
melanocytes, 30
membranous labyrinth, 274
meniges, 55-57, 173-174
menisci, 250
mesometrium, 234
mesosalpinx, 234
mesovarium, 234
metacarpophalangeal (MCP) joints, 186
metatarsal bones, 247
microglia, 22
microscope, invention, 7
microscopic anatomy, 7-8
microvilli, 13
mid-sagittal plane, 152
midbrain, brain stem, 52
middle circular layer, stomach, 113
middle layer, eyeball, 264
mineralized cartilage, 38
minor calyces, 124
mitral valve, 81
mons pubis, 144
motor areas, lobes, 48
motor pathways, central nervous system, 49-50
mouth, 99, 109-110, 268-271
movement (anatomical), 154-157

MRI (magnetic resonance imaging), 8
MS (multiple sclerosis), 22
mucin, 13
multiple sclerosis (MS), 22
muscle fiber, 17
muscles
anterior abdominal wall, 217
chewing, 269
cremasteric muscle, 134
deep back muscles, 172-173
face, 261-262
involuntary muscles, 19
pelvic region, 229
scalp, 261-262
stomach, 113
superficial back muscles, 172
upper limb, 186
arm, 188-189
carpal tunnel, 194-195
forearm, 189-192
hand, 192-194
shoulder region, 186-188
voluntary muscles, 18
muscular tissue, 17-19
musculocutaneous nerve, 198
myocardium layer, heart, 80
myometrium, uterus, 143

N

nail beds, 32
nail grooves, 32
nails, 32
nasal cavity, 98-99, 266-268
nasal septum, 266
nasopharynx, 100
neck, 274-275
nephron, kidneys, 126-127
nerve plexuses, 69
nerves, 61-63, 134
cranial nerves, 67-68, 260
intercostal nerves, 206

spinal nerves, 63-67, 173-174
   ANS (autonomic nervous system), 65-67
   SNS (somatic nervous system), 64-65
subcostal nerve, 206
nerve supply
   lower limb, 253-254
   upper limb, 197-199
nervous system, 45, 61, 68
   ANS (autonomic nervous system), 61-71
      parasympathetic system, 69-71
      sympathetic system, 69-70
   ascending pathways, 49-50
   brain, 45-46
      brain stem, 52-53
      cerebellum, 52
      cerebrum, 47-51
      diencephalon, 51-52
      sagittal view, 46
   descending pathways, 49-50
   meniges, 55-57
   nerves, 61-63
      cranial nerves, 67-68
      spinal nerves, 63-67
   SNS (somatic nervous system), 61-62, 68-69
   spinal cord, 54-55
   ventricular system, 58-59
nervous tissue, 19-22
neuroanatomy, 8
neurocranium, 258
neurons, 20-21, 64
neurotransmission, 21
neurotransmitters, 21
nodes, lymph nodes, 90-93
nose, 98-99, 266-268
nuclei, 47
nucleus pulposus, 169

O

obturator nerve, 253
occipital lobes, cerebrum, 48
oculomotor nerve (CN III), 264
olecranon process, 183
olfactory epithelium, 267
olfactory nerve (CN I), 67, 267
oligodendrocytes, 22
oocytes, 140
oogenesis, 141
openings
   pelvic region, 229
   skull, 257-259
   thoracic cavity, 205
oral cavity, 99
orbicularis oculi, 261
orbicularis oris, 262
orbit, 262-265
oropharynx, 100
os coxae, 239
Osgood-Schlatter's disease, 245
ossification, 37
osteoblasts, 15, 36
osteoclasts, 36
osteocytes, 36
osteocytic lacunae, 36
osteoids, 36
outer cortex, kidneys, 124
outer fibrous layer, eyeball, 264
outer layer of longitudinal muscle, stomach, 113
outer wall, orbit, 263
ova, 140
ovaries, 140-141, 234
oviducts, 141

P

palate, 110
palatoglossus, 270
Palmar interossei ADduct the fingers (PAD), 193
pampiniform plexus of veins, 134

pancreas, 119
   functions, 223
paramedian plane, 152
paranasal sinuses, 267
parasagittal plane, 152
parasympathetic division (ANS), 63, 69-71
parathyroid glands, 275
parietal, 218
parietal lobes, cerebrum, 48
parieto-occipital sulcus, 48
Parkinson's disease, 51
parotid, 270
parotid gland, 110
patella, 243
pectoralis major, 187
pectoral region, 202-203
pedicles, 163
pelvic diaphragm, 229
pelvic inlet, 229
pelvic outlet, 229
pelvic region, 227
   bones, 227-228
   muscles, 229
   openings, 229
   organs, 229-234
   perineum, 235-237
penis, 136-137
perforating veins, 197
pericardial cavity, 207
pericardium, heart, 207
perineum, 235-237
periosteum, 38
peripheral nervous system (PNS). *See* PNS (peripheral nervous system)
peripheral process, 64
peritoneal sac, 218-219
PET (positron emission tomography) scans, 8
Peyer's patches, 94
phalanges, 184, 247
pharynx, 99-100, 111-112, 275
phrenic nerve, 205
physiological sphincter, stomach, 113

pia mater, meniges, 55
pits, stomach, 113
planes (anatomical), 152-153
plantar flexion versus dorsiflex-
    ion, 156
plantar surface, foot, 248
pleura, 209-211
pleurisy, 210
pleuritis, 210
plexuses, 69
plexusses, 69
plicae circulares, 114
PNS (peripheral nervous
    system), 61, 68
    ANS (autonomic nervous
        system), 61-71
    nerves, 61-63
        cranial nerves, 67-68
        spinal nerves, 63-67
    SNS (somatic nervous
        system), 61-62, 68-69
podocytes, 127
pons, brain stem, 53
popliteal fossa, 252
portal venous system, 87, 120
position (anatomical), 151-153
positron emission tomography
    (PET scans), 8
postcapillary venules, 86
postcentral gyrus, 48
posterior abdominal wall,
    abdomen, 224-225
posterior compartment, 188,
    243-244, 247
posterior cruciate ligaments, 250
posterior lobe, prostate gland,
    232
posterior longitudinal ligament,
    170
posterior parts, 152
posterior view, pelvis, 228
potential space, 209
precentral gyrus, 48
premature babies, lungs, 105
prepuce, penis, 136
prickle cell layer, epidermis, 29

primary curves, spine, 160
primary trabeculae, 38
processes, vertebrae, 163
promontory of the sacrum, 229
pronation versus supination, 155
prostate gland, 135-136
    lobes, 232
prostatic urethra, 134, 232
proximal radioulnar joint, 184
proximal versus distal, 153
pseudostratified epithelial tissue,
    13
psoriasis, 30
pubic crest, 229
pubic symphysis, 229
pubis, 228
pulled hamstrings, 245
pulmonary circulation, 82-83
pyloric sphincter, stomach, 113

Q-R

quadrants, breast region, 203

radial artery, 196
radial groove, 181
radial nerve, 197
radial notch, 183
radial tuberosity, 183
radiocarpal joint, 184
radioulnar articulation, 183
radioulnar joints, 183-184
radius, 183
rami, 64
rami communicantes, 66
rectum, 231
rectus abdominis, 217
reflective media, eyeball, 265
regions
    spine, 160
    vertebrae, 164-169
relationships (anatomical),
    152-153
renal, 124
renal capsule, 124

renal columns, 124
renal corpuscle, 126
renal papilla, 124
renal pyramids, 124
renal sinus, 124
reproductive system
    female reproductive system,
        139-140
        external genitalia, 144-145
        mammaries, 145-147
        ovaries, 140-141
        uterine tubes, 141-142
        uterus, 142-143
        vagina, 144
    male reproductive system,
        131
        accessory glands, 135-136
        ductus deferens, 133-134
        epididymis, 133-134
        penis, 136-137
        scrotum, 131-132
        testicles, 131-133
respiration, 105
respiratory alveoli, 105
respiratory bronchiole, 104-105
respiratory diaphragm, 205
respiratory epithelial tissue. *See*
    pseudostratified tissue
respiratory system, 97
    gaseous exchange, 107
    lower respiratory tree, 102
        bronchi, 103-104
        lungs, 104-106
        trachea, 103-104
    upper respiratory tree, 97-98
        larynx, 100-102
        nasal cavity, 98-99
        oral cavity, 99
        pharynx, 99-100
reticular fibers, connective
    tissue, 15
retina, eyeball, 264
ribs, 205-206
ridges, stomach, 113
right margin of the heart, 208
rotator cuff, 187-188

rugae, 113
ruptures, rotator cuff, 188

S

sacculations, 221
sacral canal, 167
sacral hiatus, 167
sacroiliac joint, 167
sacrotuberous ligaments, 229
sacrum, 167-168
sagittal plane, 152
sagittal view
    brain, 46
    female pelvic organs, 233
    male pelvic organs, 231
saliva, 111
salivary glands, 110-111,
    268-271
scalp muscles, 261-262
scapula, 178-179
scapulothoracic joint, 181
sciatic nerve, 253
scoliosis, 161-162
scrotal wall, 132
scrotum, 131
sebaceous glands, 32-33
secondary bronchi, 103
secondary curves, spine, 160
secondary trabeculae of bone, 38
segmental bronchi, 103
semilunar valves, 81
seminal vesicles, 134-135
seminiferous tubules, testicles,
    133
senses
    hearing, 272-274
    smell, 266-268
    taste, 268-271
    vision, 262-265
sensory areas, lobes, 48
sensory neurons, 64
sensory pathways, central
    nervous system, 49-50

serous pericardium, 207
serratus anterior muscle, 187
sex hormone production,
    testicles, 133
short bones, 38
shoulder region
    bones, 178-181
    joints, 181-182
    muscles, 186-188
simple columnar epithelia, 13
simple cuboidal epithelia, 13
simple epithelial tissue, 12-13
single midline third and fourth
    ventricles, brain, 58
skeletal cells, 15
skeletal muscle, 18
skeletal system, 35, 39
    appendicular system, 40-41
    axial system, 39-40
    bones, 35-36
        cells, 36
        growth and development,
            36-39
    joints, 41
        nonsynovial joints, 41-42
        synovial joints, 42-43
skin, 28-31
skull, 257-259
small intestine, 114-117, 220
smell (sense of), 266-268
smooth muscle, 19
SNS (somatic nervous system),
    61-62, 68-69
    spinal nerves, 64-65
solid joints, 41-42
somatic motor, spinal nerves, 67
somatic nervous system (SNS),
    61-62
somatic sensory, spinal nerves,
    67
spermatic cord, 134
spermatogenesis, 133
sphenoid sinus, 266
sphincters, stomach, 112-113

spinal cord, 54-55, 173-174
    CSF (cerebrospinal fluid),
        56-59
    dura sac, 56
spinal nerves, 63-67, 173-174
spine, 159
    curves, 160-162
    regions, 160
    vertebrae, regions, 163-169
spinosum stratum, epidermis, 29
spinous process, vertebrae, 163
spiral groove, 181
spleen, 94, 220
stapedius, ear, 273
stapes, ear, 273
sternoclavicular joint, 181
sternum, 178
stomach, 112-113, 220
stratified squamos epithelial tis-
    sue, 13
stratum basale, epidermis, 29
stratum corneum, epidermis, 28
stratum granulosum, epidermis,
    29
stratum lucidum, epidermis, 28
stratum spinosum, epidermis, 29
striated muscle, 18
structures, locations, 153
styloglossus, 270
subarachnoid space, 55-56
subcostal nerve, 206
sublingual gland, 110, 271
submandibular gland, 110, 270
subscapularis, 188
sulci, brain, 47
superficial back muscles, 172
superficial veins, 196
superficial versus deep, 153
superior border of the heart, 209
superior pole, kidneys, 123
superior thoracic aperture, 205
superior vena cava, 86
superior wall, orbit, 263
supination versus pronation, 155
suprarenal glands, 123

supraspinatus, 188
supraspinous ligament, 170
surfactant, 105
sutural joints, 41
swallowing, 111
sweat glands, 32-33
sympathetic division (ANS), 63
sympathetic system, ANS (autonomic nervous system), 69-70
symphysis joints, 42
synapses, 20
synaptic clefts, 21
synarthroses joints, 41-42
synchondrosis, 42
syndesmoses, 41
synovial fluid, 43
synovial membranes, 43
systemic arteries, 83
systemic circulation, 82-83
systole, 81

T

T-lymphocytes, 91
tarsal bones, 38, 247
taste (sense of), 268-271
tears, rotator cuff, 188
temporal bone of the skull, 257
temporal lobes, cerebrum, 48
temporomandibular, 257
teniae coli, 221
tensor fasciae latae, 241
tensor tympani, ear, 273
teres minor, 188
tertiary bronchi, 103
testicles, 131-133
testicular artery, 134
thalamus, diencephalon, 51
thigh, 242-245
thin medial walls, orbit, 263
third and fourth ventricles, brain, 58
thoracic cavity, 206
thoracic duct, 92

thoracic organs, 206
  heart, 207-209
  lungs, 209-211
    breathing mechanics, 212-213
  pleura, 209-211
thoracic vertebrae, 166
thoracic wall, 204-206
thoracolumbar outflow, sympathetic division, 70
throat, 99-100
thymus, 94-95
thyroid gland, 275
tibial nerve, 254
tibial tuberosity, 245
tissue, 11
  bones, 36
  connective tissue, 14-17
  epithelial tissue, 11-14
  muscular tissue, 17-19
  nervous tissue, 19-22
TMJ (jaw joint), 257
toenails, 32
tongue, 110, 268-271
tonsils, 94
trabeculae, 38
trabecular bones, 39
trachea, 103-104, 275
transitional epithelial tissue, 14
transverse foramina, 165
transverse plane, 152
transverse processes, vertebrae, 163
transversus abdominis, 217
trapezius, 187
trigone, 230
trochlea, 183
trochlear nerve (CN IV), 67, 264
trochlear notch, 183
true pelvis, 229
tubules, 133
tunica albuginea, 140
  testicles, 133

U

UG (urogenital) triangle, 235-236
ulna, 183
ulnar artery, 196
ulnar nerve, 199
undersurface, skull, 259
upper inner quadrant (breast), 203
upper limb, 177
  bones, 178-185
  joints, 178-186
  muscles, 186-194
upper outer (breast), 203
upper respiratory tree, 97-98
  larynx, 100-102
  nasal cavity, 98-99
  oral cavity, 99
  pharynx, 99-100
upper surface, kidneys, 123
ureter, 124
ureters, 127, 225-231
urethra, 128-129
urethral crest, 232
urinary bladder, 127, 231
urinary system, 123
  kidneys, 123-125
  ureters, 127
  urethra, 128-129
  urinary bladder, 128
urine, contents, 127
urogenital (UG) triangle, 235-236
urogenital hiatus, 229
uterers, males, 230
uterine tubes, 141-142
uterus, 142-143

V

vagina, 144
valves, heart, 80-82
Van Leeuwenhoek, Antoine, 7
vas deferens, 134

vasectomies, 134
veins, 86
    anal canals, 237
    azygous system of veins, 206
    inferior vena cava, 86
    lower limb, 253
    perforating veins, 197
    portal venous system, 87, 120
    superficial veins, 196
    superior vena cava, 86
ventral primary ramus, 65
ventricles, brain, 58
ventricular diastole, 81
ventricular system, 58-59
ventricular systole, 81
vermis, 52
vertebra prominens, 165
vertebrae, 159, 163-164
    cervical vertebrae, 164-165
    coccyx, 169
    lumbar vertebrae, 166-167
    regions, 164-169
    sacrum, 167-168
    thoracic vertebrae, 166
vertebral arches, 163
vertebral column, 159
    ligaments, 170-171
    vertebral joint, 169-170
vertebral foramen, 163
vertebral joint, 169-170
vertebral notches, 164
Vesalius, Andreas, 6-7
vessels, 91
    blood vessels, 84-85
    capillaries, 85
    heart, 82-83
    lymph trunks, 91
    systemic arteries, 83
    thoracic duct, 92
vestibular folds, 101
vestibular nerve, 274
vestibulocochlear nerve (CN
  VIII), 274

visceral motor, spinal nerves, 67
villi, 114
visceral layer, 218
visceral pleura, 209
viscerocranium, 258
vision, 262-265
vitreous body, eyeball, 265
vocal cords, 102
voice box, 100-102
voluntary muscles, 18

W–X–Y–Z

walls
    orbit, 263
    small intestine, 115
white blood cells, connective
  tissue, 15
white matter, cerebrum, 47
windpipe, 103-104
wrist, 184-185, 194-195